POLYOMINOES
A Guide to Puzzles and Problems in Tiling

George E. Martin

The
Mathematical
Association
of America

POLYOMINOES
A GUIDE TO PUZZLES AND
PROBLEMS IN TILING

GEORGE E. MARTIN

MAA
SPECTRUM

**To
Margaret**

BP 1   94/95

Foreword

This little book of puzzles and problems can be your guide to entertainment for years. A glance at the illustrations as seen in a flip of the pages tells more about the subject of these puzzles and problems than words could manage. No formal mathematical training is required to enjoy this book. In fact, the things taught in school will be of small value here. The puzzles and problems, which for simplicity are all labeled problems in the text, present a wide range of difficulty. Some require only patience; some require more patience than most of us can muster; some require skill and insight; and some require cleverness that has yet to be exhibited by anyone. Indeed, some of the problems have yet to be solved. (At present, unsolved problems include those that are numbered 1.4, 3.11, 3.12, 4.13, 4.14, 4.15, 5.4, 5.22, 5.27, 6.10, 7.11, 7.14, 8.13, 8.15, 8.16, 9.9, 9.16, 9.17, 9.18, 9.19, and 9.20 in the text.) Therefore, it is impossible to say how long it would take to work your way through the book, page by page, attempting every puzzle that you encounter. More likely you will want to skip many of the puzzles, and you are encouraged to do so. You should work on the puzzles and problems that interest you.

Even if you read through the entire book and ignore the frequent requests that ask you to stop and try some of the problems as you go along, you should still find a lot of things that are interesting to you. It is possible that the best way for you to enjoy this book is to read rather quickly the entire book before returning to the puzzles and problems that attract your special interest. It will not matter that in your reading you have seen the many solutions that appear. Unless you have purposely studied these solutions, the corresponding problems will be as challenging as before. This is somewhat analogous to having previously seen the picture of a jigsaw puzzle that you are about to put together. The analogy carries further in that the many puzzles that are like jigsaw puzzles are as enjoyable and almost as challenging when they are attempted again at a later date.

The problems are of such a varied nature that you are bound to find many to your liking. You should feel free to skip about. If you see a word you don't understand, the index will show you where its meaning can be found. One more thing, you should be warned that playing with polyominoes can be habit forming.

Special thanks go to Richard Collier, Margaret Farrell, Lynn Green, Victoria Kouba, and John Malkevitch, who gave many helpful suggestions after reading a preliminary version of this book.

> George E. Martin
> Albany, New York
> July 1991

CONTENTS

---

* "Pentominoes" is a registered trademark of Solomon W. Golomb.

THE MONOMINO □
AND OTHER POLYOMINOES

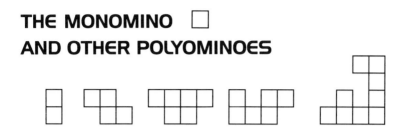

Let's begin with a problem. For our purposes, a *domino* is only a
1 × 2 rectangle. Given an 8 × 8 board, it is not too difficult to cover this
board with exactly 32 dominoes, since each domino covers exactly two of
the sixty-four squares of the board. Therefore, to make a more interesting problem, suppose that two diagonally opposite corner squares are
removed or otherwise covered. Can you cover the remaining 62 squares
of the board with 31 dominoes? Figure 1.1 shows the board and one
domino. A domino can be turned around but must be placed on the
board to cover exactly two squares. Try to solve this problem before we
get back to it.

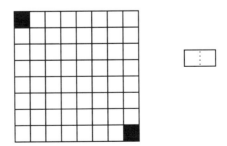

FIGURE 1.1

If you happen to have a set of dominoes that was purchased for playing the game by the same name, then you will find that one set is not enough for the problem above. One face of a standard domino that you purchase is divided into two squares, each marked with anywhere from 0 to 6 dots. There are no duplications; the 3&5 domino is the same as the 5&3 domino. How many dominoes are there in a standard set? Can you compute the number without making an exhaustive list? You will have to give special consideration to the doubles, which are the dominoes of the form $m\&m$. Depending on how you take the doubles into consideration, the computation for the desired number is probably $[7] + (1/2)[(7 * 7) - 7]$ or $[(7 * 7) + 7]/2$ or $7 + 6 + 5 + 4 + 3 + 2 + 1$. From now on, we do not care about the markings on a domino; for us, a domino is simply a 1×2 rectangle. Think of a domino as two unit squares that are joined along a common edge. As we will see, polyominoes are a generalization of dominoes. (We will build many words with the root - omino. The pronunciation of *polyomino* begins as does *polygon* and ends as does *domino.*)

You will enjoy polyominoes more if you do the problems as you go along. Consider this book as a guide to your study and enjoyment. Don't just read about the problems. It is generally more fun to do things than to read about them, isn't it? You should read until you come to an interesting problem. Then put the book aside and work on that problem. When you have exhausted the problem—or perhaps when the problem has exhausted you—return to the book. Solutions will be provided for many of our problems, while some problems will have only hints. There will even be some problems for which solutions are not known.

Other than graph paper, you do not need extensive materials for the problems posed in this book. The polygons needed can be cut from lightweight cardboard. Manila file folders provide nice stock. Ordinary paper is too thin for much handling as the edges tend to curl. On the other hand, ordinary scrap paper is readily available and used pieces can be discarded and replaced with ease. Draw the desired polygons on a sheet of graph paper, place the graph paper on about eight pieces of scrap paper, staple the pile in several places to prevent sliding, and cut out many copies of the desired polygon at once. Using two centimeters or else one inch as the unit distance is about right for scale. However, for the domino problems, if you happen to have a checkerboard handy, you can cut your dominoes to fit over two squares of the checkerboard.

The suggestion to use a standard checkerboard for our first problem contains within it the germ of a solution to the problem. We now restate the problem above in an easier form and then generalize the problem. It often happens that consideration of a generalized problem helps to solve the specific problem. See Figure 1.2.

Problem 1.1.

A domino is a rectangle that covers exactly two adjacent squares of an 8 × 8 checkerboard. If two diagonally opposite corner squares of the checkerboard are removed or otherwise covered, can you cover the remaining 62 squares with 31 dominoes? Suppose the two squares to be removed are picked at random from the 8 × 8 checkerboard; can the remaining squares be covered with 31 dominoes in this case?

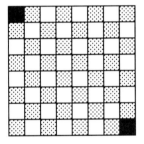

FIGURE 1.2

Mathematics is one subject where we can sometimes prove something cannot be done. For example, in more advanced mathematics it can be proved that an angle of 60° cannot be trisected with a ruler and compass. Yet, every so often someone appears with a construction that is claimed to have been overlooked by mathematicians for two thousand years. There is always a flaw in the argument. This is necessarily so since it has been proved once and for all that there can be no solution. There is a considerable difference between knowing a solution does not exist and not knowing a solution. If you have been unable to

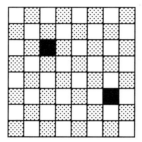

FIGURE 1.3

find a solution to the checkerboard problem indicated in Figure 1.2, it is because there is no solution. The situation is quite different with the checkerboard problem indicated in Figure 1.3. What about the general case of two randomly chosen squares removed from the board? We will return to these questions later, but you should try them for yourself at this time.

Polyominoes

In the game of chess, the rook is allowed to move only horizontally and vertically. A set of squares is said to be rookwise connected if you can get from any square to any other square by a sequence of rook moves without leaving the set. See the examples accompanying the chapter title. As a *domino* is made of two rookwise connected squares, so a *tromino* is made of three rookwise connected squares. Hence, a tromino consists of three squares, each attached along at least one edge to at least one of the other two squares. (Trominoes are often called triominoes.) We allow our polygons to be rotated and to be flipped over. Therefore, there are only two trominoes, the straight and the *L*. See Figure 1.4. A *tetromino* is made of four rookwise connected squares. So a tetromino consists of four squares, each attached along at least one edge to at least one of the other squares. Players of the computer game TETRIS™ are familiar with tetrominoes. There are five tetrominoes, the straight, the square, the *L*, the *T*, and the *Z*. The *Z*-tetromino is also called the skew tetromino. If you think a *pentomino* is made of five rookwise con-

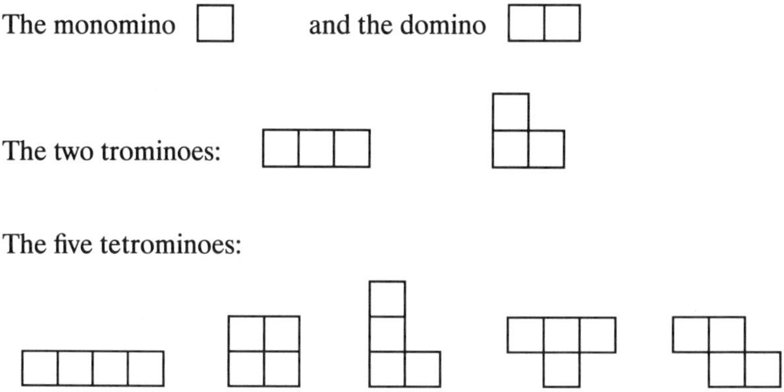

The monomino ☐   and the domino ☐☐

The two trominoes: ☐☐☐   [L-shaped tromino]

The five tetrominoes:

[Five tetromino figures]

FIGURE I.4

nected squares, you are correct and have anticipated the next numbered problem.

Problem I.2

How many pentominoes are there? Draw a set of the pentominoes.

For positive integer n, an *n-omino* is defined to be the union of a set of n unit squares from an infinite checkerboard such that any two of these squares are connected by a finite number of rook moves within the set. All the n-ominoes are called *polyominoes.* One way to picture the polyominoes is that they are the pieces that can be torn from an arbitrarily large sheet of square postage stamps. Two polygons are considered different only if they are not congruent. Note that not every union of unit squares is a polyomino. In a polyomino, adjacent squares are joined edge-to-edge. See Figure 1.5 for four examples that are not polyominoes. A 1-omino is a unit square and is called a *monomino*. So a 1×1 square, a unit square, a 1-omino, and a monomino are all the same thing. (Why shouldn't one thing have several names? You do, depending on who is talking to you.)

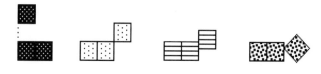

FIGURE 1.5

Solomon W. Golomb introduced the word "polyomino" in his 1954 paper "Checker Boards and Polyominoes" in *The American Mathematical Monthly* (volume 61, pp. 675–682). This paper is the published version of a talk Golomb gave to the Harvard Mathematics Club in 1953. A larger public was exposed to polyominoes in May and December of 1957, when Martin Gardner cited Golomb's work in his column "Mathematical Games" in the journal *Scientific American*. This was the beginning of a sequence of articles in Gardner's column that were to deal with polyominoes and related subjects. The subject was summarized by Golomb in his book *Polyominoes* (Scribner's, 1965). In this famous book, Golomb tells us that the first pentomino problem was published in 1907 in the classic puzzle book *Canterbury Puzzles* by H. E. Dudeney and also that he had learned that further literature on the subject of polyominoes appeared beginning in the 1930s under the heading of "dissection problems" in the British puzzle journal *Fairy Chess Review*. However, since the terminology used today and the present interest in the subject began with Golomb's study, he is rightfully called the father of polyominoes.

From one point of view there is a lot to say about the monomino. One question about unit squares is the following. Given a square, how many unit squares can you pack into the given square? We never allow our figures to overlap except along corners and edges. Cutting a unit square into pieces is not allowed. If the length of a side of the given square is an integer n, then the problem is too easy and has the answer n^2. However, how many unit squares can you pack into a square with side 2.75? The answer happens to be 5. Figure 1.6 shows five unit squares packed into each of two squares. These two squares have sides of different lengths. Which packing leads to an answer to the question just asked is left for you to figure out on your own. Well, you say, 0.75 is large enough relative to the side of a unit square so that it is no big sur-

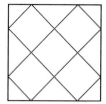 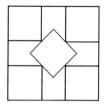

FIGURE 1.6

prise that more than four unit squares can be squeezed into the square with side 2.75. What happens when the side is only a little bigger than an integer? Most people feel that if a square has sides of length s where $s = n + (1/10)$, then, even if n is a large integer, there is about $s/5$ square units of wasted space after the obvious n^2 unit squares have been stacked in one corner. What possible good could it do to place some of the unit squares at various odd angles? It has got to be surprising that for large values of n it does, in fact, do some good. With $n = 100,000$ and $s = 100,000.1$ there will be room for at least 6400 more unit squares, in addition to the expected n^2. Martin Gardner's October 1979 column reports on this amazing result, which is due to Paul Erdős and Ronald L. Graham, along with other results on packing problems.

From another point of view there is little to say about the monomino. However interesting the packing problems mentioned in the paragraph above are, they are not really polyomino problems. Most people in the study of polyominoes draw the line at figures that do not conform to the infinite checkerboard. So all the vertices of a polyomino sitting in the coordinate plane have integers for coordinates. We will take this general point of view, and this leaves little more that can be said about monominoes.

Coloring Boards

Let's return to our first problem, which is much easier when stated as an 8×8 checkerboard problem rather than merely as an 8×8 board problem. The coloring of the checkerboard is the key to showing that it is

impossible to cover the remaining squares with dominoes after two diagonally opposite corner squares are removed, as in Figure 1.2. Strangely
enough, in mathematics checkerboards are usually colored black and
white. This convention makes little difference to us; what is important is
the alternating color pattern of some two colors—any two colors. When
a domino is placed on a checkerboard, the domino covers one black
square and one white square. Thus any set of squares that are covered
by dominoes must contain the same number of black squares as white
squares. In particular, the 31 dominoes needed to cover the 62 squares
would have to cover 31 black squares and 31 white squares. However,
diagonal corners of an 8×8 board have the same color. In Figure 1.2,
we have 30 white squares and 32 black squares to cover. It is impossible to cover 30 white squares and 32 black squares with 31 dominoes.
Therefore, we know that the first part of Problem 1.1 is impossible. This
is very different from not knowing a solution for the problem. When you
know there is no solution, you stop looking for one.

　　From the fact that a domino covers one square of each of the two
colors of a checkerboard, we conclude that if two squares are to be removed from an 8×8 checkerboard such that the remaining squares are
to be covered with dominoes then the two squares must be of different colors. If 62 squares of a checkerboard are covered with 31 dominoes, then it is necessary that half of the 62 squares are black and half
are white. Surprisingly, it turns out that this condition is also sufficient.
That is, any 31 black and 31 white squares from an 8×8 checkerboard
can be covered with 31 dominoes. To see this, consider Figure 1.7. The

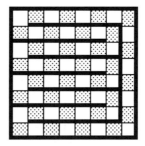

FIGURE 1.7

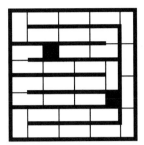

FIGURE 1.8

board there is partitioned by the heavy line segments to form a path that covers every square and returns back on itself, where the colors of the squares along the path alternate. Hence if one square of each color is removed, each section of the path between the removed squares can be covered with dominoes. Of course, it is important that dominoes can turn corners along the path. Following this procedure, we have Figure 1.8 as one solution to the problem of Figure 1.3. What can you say about covering a 7×7 checkerboard?

Problem 1.3

Can 24 dominoes and 1 monomino cover a 7×7 checkerboard?

There are 12 pentominoes. They are conventionally named after letters of the alphabet. See Figure 1.9 and remember

$$FLIP \quad 'N \quad TUVWXYZ.$$

Noting that the last seven letters of the alphabet are used, you should have no trouble with the names, once you have stretched your imagination to accept the N.

Let $p(n)$ denote the number of n-ominoes. We already know that $p(1) = p(2) = 1, p(3) = 2, p(4) = 5$, and $p(5) = 12$. There is an obvious challenge.

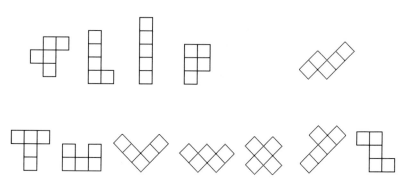

FIGURE 1.9

Problem 1.4

Give a formula for the number of n-ominoes for positive integers n.

Although this is an obvious polyomino problem to state, a solution is not obvious. In fact, there is no known formula for $p(n)$. Unlike the situation for Problem 1.1, we are not able to prove that there is no nice formula. Here we have a problem for which we simply do not know the answer. The important difference between not knowing a solution to a problem and knowing there is no solution to that problem is sometimes confused. Near the end of the book we will mention values of $p(n)$ for other small n, in case you would like to have a go at finding some of these values on your own. How many *hexominoes,* which are the 6-ominoes, are there?

There is one more thing to mention concerning the definition of a polyomino. The problem does not appear until we get to the *heptominoes,* which are the 7-ominoes. Is the union of unit squares indicated in Figure 1.10 a polyomino? What do you think? Should polyominoes be allowed to have holes? The world of those confronted with this question is divided. The answer to the question affects the value of $p(7)$. Until something comes along to convince one group that they should yield to the other in this matter of formulating the definition—something such

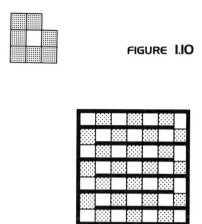

FIGURE 1.10

FIGURE 1.11

as a nice formula for $p(n)$ in one case but not the other—there will re-
main two groups. For our purposes, you are free to join either group.

The path indicated in Figure 1.11 gives an answer to our Problem
1.3. A 7×7 checkerboard with white on the corners can be covered with
24 dominoes and one monomino if and only if the monomino is placed
on a white square. The monomino must be placed on a white square because the dominoes cover an equal number of black and white squares
and there is 1 more white square than black squares on this checkerboard. After a white square has been covered with the monomino, the
remaining paths in Figure 1.11 are easily covered with dominoes. What
is the story for a 9×9 checkerboard with one square removed? What
is the story for a 10×10 checkerboard with two squares removed? By
now, you should be able to give solutions to the problems that are the
generalizations of Problems 1.1 and 1.3 to $n \times n$ checkerboards. You
may even feel confident enough to tackle rectangular checkerboards.

You will notice that some problems are numbered for emphasis
and for reference. Many problems are mentioned only in passing and
their solutions are not given here. However, you should not hesitate
to work on an unnumbered problem that attracts your attention. You
should work on the problems that interest you. We will end this section
with a hard problem, but one that requires no techniques beyond those
you have already seen.

Problem 1.5

Find the 35 hexominoes and show that they cannot be put together to form a rectangle.

THE DOMINO   ⊓⊔

A *copy* of a polygon or polyomino P is a figure that is congruent to P. So a copy of P may be rotated, flipped over, or slid around in re-lation to P. A region R, such as a rectangle, a polyomino, a quadrant, or the plane, is *tiled with* P if R is covered with copies of P that do not overlap except along their edges. As one might expect, these copies of P in the tiling are called *tiles*. This terminology should be fairly familiar from everyday life. It is the terminology that will be used here and gen-eralizations of this terminology to tilings with more than one polygon or polyomino should not cause any confusion. Of course, we don't actually encounter a tiling of the (infinite) plane in everyday life. However, we are quite used to seeing a finite pattern that we realize can be extended arbitrarily far in any direction. We associate such a pattern with a tiling of the plane. A tiling is also sometimes called a mosaic, a paving, or a tessellation.

We return to the domino, which dominated our first section. The first of two new ideas to be introduced here concerns tilings in which you cannot pick out a proper rectangle in the tiling. Specifically, a tiling with dominoes or with any set of rectangles is said to be *simple* unless a subset of at least two but not all of the tiles form a rectangle in the tiling or else the tiling contains only one or two tiles. The second condition eliminates the trivial cases of a rectangle tiling itself and of a rectangle split into only two rectangles. The tiling of the rectangle with nine dominoes in Figure 2.1 is certainly not simple, since you should be able to pick out thirteen smaller rectangles by making certain selections of two to six of the dominoes in the tiling.

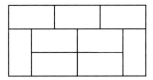

FIGURE 2.1

A *fault line* for a tiling is a straight line that intersects the interior of the tiling but does not intersect the interior of any tile in the tiling. The tiling of the rectangle in Figure 2.1 has one fault line. This fault line is horizontal and separates the top three dominoes from the bottom six. A tiling containing at least two tiles but without a fault line is said to be *fault-free*. So a rectangle tiled with more than one domino is fault-free if the tiling cannot be split into two tiled subrectangles.

We will discuss the first five of the following problems in some detail. As usual, you are urged to try these problems before you read about their solutions.

Problem 2.1

Show that no rectangle has a simple tiling with dominoes.

Problem 2.2

Find the simple tiling of the plane with dominoes.

Problem 2.3

Show that if an $h \times w$ rectangle has a fault-free tiling with dominoes, then $h \geq 5$ and $w \geq 5$.

Problem 2.4

Find fault-free tilings with dominoes for each of a 5 × 6 rectangle and a 6 × 8 rectangle.

Problem 2.5

Can you find a fault-free tiling of a 6×6 square with dominoes?

Problem 2.6

Extend a fault-free tiling with dominoes of a 5 × 6 rectangle to a fault-free tiling with dominoes of an 8 × 8 square.

Problem 2.7

Find a fault-free tiling of the plane with dominoes that is not simple.

Simple Tilings

In attempting to start a simple tiling with dominoes within a right angle of a rectangle, we find that once the tile labeled 1 is placed in Figure 2.2, then tiles labeled 2, 3, and 4 are determined in order and that the attempt has failed as the tiles labeled 1 and 4 form a rectangle . Therefore, we immediately conclude that no rectangle can have a simple tiling with dominoes. Can you convince yourself that no simple tiling of the plane with dominoes can have a fault line?

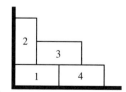

FIGURE 2.2

FIGURE 2.3

We now attempt to start a simple tiling of the plane with dominoes. An edge of length 2 of a domino in such a tiling must intersect an
edge of length 1 of a neighboring domino, since otherwise a 2 × 2 or a
1 × 4 rectangle formed by two dominoes would be forced as in Figure
2.3. Therefore, considering symmetry, we can suppose that we start with
one of the three general positions indicated by the black dominoes in
Figure 2.4. The first and third cases quickly lead to the indicated fail-
ures, since the spaces starred in the figure cannot be filled without forming a rectangle. The special, second case is thus the only one permitted.
Beginning with the two dominoes of this special case, we see that all
their neighbors are uniquely determined. These neighbors are shown
in Figure 2.4. Likewise, the neighbors of neighbors are always uniquely
determined in the same way by the special case. The special case then
does not lead to failure but determines the unique simple tiling of the
plane with dominoes. This familiar tiling is indicated in Figure 2.5. Unlike the tiling of the rectangle in Figure 2.1, there is no line along which
this tiling folds on itself. However, note that if you turn over a copy of
the tiling, you can slide the copy to fit exactly over the original. This
is not obvious. One way of checking this claim is to hold two copies of
Figure 2.5 facing each other in front of a strong light.

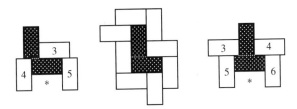

FIGURE 2.4

Fault-free Tilings

The unique simple tiling of the plane with dominoes, which is indicated in Figure 2.5, is fault-free. The rectangle tiled with dominoes in Figure 2.6 has one fault line. This fault line divides the tiled rectangle into two tiled rectangles. Check that the tilings of each of these two rectangles, one 5×6 and the other 6×8, is fault-free. That solves Problem 2.4. As we will see, the answer to Problem 2.5 is No.

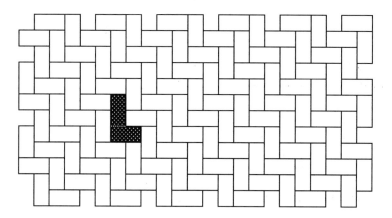

FIGURE 2.5

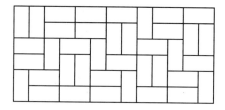

FIGURE 2.6

We are going to end this chapter by following an argument of R. L. Graham that determines exactly which rectangles have a fault-free tiling with dominoes. You can skip the proof without risking later understanding. If you do, however, you will miss a nice argument that proves a nice result.

Suppose that we have a rectangle with height h and width w and suppose that the rectangle has a fault-free tiling with dominoes. We will show below that $h \geq 5$. Since an $h \times w$ rectangle can also considered to be a $w \times h$ rectangle, it will then follow by symmetry that we must also have $w \geq 5$. Note that hw must be even, whether the tiling is fault-free or not. This is a consequence of the fact that each domino covers two unit squares. Since hw is even, at least one of h or w must be even.

A $2 \times w$ rectangle tiled with dominoes necessarily has a fault line for $w > 1$. That a $3 \times w$ rectangle tiled with dominoes also necessarily has a fault line follows from the same kind of argument that we will apply below to a $4 \times w$ rectangle. The $3 \times w$ case is left for you to check.

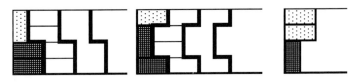

FIGURE 2.7

We will attempt to find a fault-free tiling with dominoes of a $4 \times w$ rectangle. For each of the three permissible starting positions along one end that are indicated by the black dominoes in Figure 2.7, the three dominoes touching the left border are then determined and the third case shown in the figure is seen to be the same as the first case by symmetry. Now, the placing of the next four dominoes is determined in each of the remaining two cases, as shown in the figure, if the tiling is to avoid a fault line. However, at this time we are presented with the same outline to be filled with dominoes as before. We have to add an additional four dominoes, only to be presented with the same outline to be filled again. If we do not admit to failure, we are doomed to an infinite sequence of repetitions. The infinite sequence of repetitions does tile an infinite strip but not a rectangle without a fault line. It is impossible to find a fault-free tiling of a $4 \times w$ rectangle with dominoes. It follows that if there is a fault-free tiling with dominoes of an $h \times w$ rectangle, then $h \geq 5$. Likewise, $w \geq 5$ by symmetry.

There is something more to be learned from the attempt that just failed. We might glean a method of forming larger fault-free rectangles from smaller ones. Suppose we have a fault-free tiling with dominoes of an $h \times w$ rectangle. All the dominoes along a side of the rectangle cannot be oriented the same way, as otherwise these dominoes would create a fault line. Select a domino having a side of length 2 along a side of length h of the rectangle. Pry off the selected domino to form an outline such as that darkened in Figure 2.8. Insert a "column" of h dominoes along the outline, as indicated by the black dominoes in the figure. Finally, replace the selected domino to form a new tiled rectangle having the same height but having width $w + 2$. The tiling of the new rectangle is fault-free if and only if the tiling of the original rectangle is fault-free. Hence, given a fault-free tiling with dominoes of an $h \times w$ rectangle, we can easily form a fault-free tiling with dominoes of an $h \times (w + 2)$ rectangle.

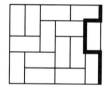 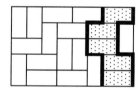

FIGURE 2.8

What works for the width above will work for the height. That is, given a fault-free tiling with dominoes of an $h \times w$ rectangle, we can easily form a fault-free tiling with dominoes of an $(h + 2) \times w$ rectangle. Also, this is one of those cases where what you can do once, you can do again and again. See Figure 2.9. (The darkened outlines in the figure are not fault lines since fault lines are straight.) From a fault-free tiling with dominoes of an $h \times w$ rectangle, we can produce, by repetition of the construction above, a fault-free tiling with dominoes of an $(h + 2m) \times (w + 2n)$ rectangle for any nonnegative integers m and n. If h

and w are both greater than 5 and exactly one of h and w is odd, then a fault-free tiling with dominoes of an $h \times w$ rectangle can be constructed from the fault-free tiling with dominoes of the 5×6 rectangle in Figure 2.8. Except for the case $h = w = 6$, if h and w are both greater than 5 and are both even, then a fault-free tiling with dominoes of an $h \times w$ rectangle can be constructed from the fault-free tiling with dominoes of the 6×8 rectangle in Figure 2.9. As we noted before, if both h and w are odd, then an $h \times w$ rectangle cannot be tiled with dominoes.

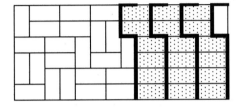

FIGURE 2.9

What about the remaining case of a 6×6 rectangle? This is another of those situations where we can prove that something cannot be done. Assume to the contrary that a 6×6 rectangle has a fault-free tiling with dominoes. Since a domino is a 1×2 rectangle, each domino in the tiling blocks exactly one of the ten potential fault lines. See Figure 2.10, where

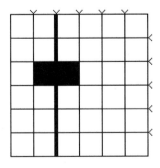

FIGURE 2.10

the ten potential fault lines are checked. We will now show that none of the ten potential fault lines can be blocked by exactly one domino. Assume some potential fault line is blocked by exactly one domino. Then, as in Figure 2.10, there remains an odd number of unit squares on each side of the line to be covered with dominoes. Of course, this is impossible. So each of the ten potential fault lines must be blocked by at least two dominoes. The ten potential fault lines then require a minimum of twenty dominoes to avoid having a fault line. This is absurd because only eighteen dominoes cover the 6×6 square. Hence, there is no fault-free tiling of a 6×6 square with dominoes.

Let's summarize in one statement our results on fault-free tilings with dominoes of rectangles. *An $h \times w$ rectangle has a fault-free tiling with dominoes if and only if*

1. *hw is even,*
2. *$h \geq 5$,*
3. *$w \geq 5$, but*
4. *not $h = w = 6$.*

THE TROMINOES

There are two trominoes, the straight tromino and the L-tromino. Which $h \times w$ rectangles can be tiled with the straight tromino? Of course, h and w must be positive integers. It is necessary that hw be divisible by 3, since a tromino covers three unit squares. That 3 divides hw is also sufficient, since then 3 divides at least one of h and w and since it is trivial to tile a $3m \times n$ rectangle with mn of the 3×1 rectangles. Therefore, an $h \times w$ rectangle can be tiled with straight trominoes if and only if 3 divides hw. The problems in the following list are more interesting and will turn up in the discussion below. Give them a try.

Problem 3.1

Tile an 8×8 checkerboard with twenty-one straight trominoes and one monomino.

Problem 3.2

Can you tile an 8×8 checkerboard with twenty-one L-trominoes and one monomino?

Problem 3.3

Can you tile a 5×5 square with eight L-trominoes and one monomino, no matter where the the monomino is placed?

Problem 3.4

Can you tile a 7×7 square with sixteen L-trominoes and one monomino, no matter where the the monomino is placed?

Problem 3.5

Can you tile a 9×9 square with twenty-seven L-trominoes?

Problem 3.6

For which integers k can you use k of the L-trominoes to tile a polyomino that is similar to the L-tromino?

Problem 3.7

Find a simple, fault-free tiling of the plane with 2×3 rectangles.

Problem 3.8

Show that there are infinitely many simple, fault-free tilings of the plane with L-trominoes.

The Straight Tromino

The first of the tromino problems above has an interesting solution. To get at this solution, we ignore the usual color pattern of the checkerboard and tricolor the board, say with the colors red, white, and blue, as indicated in Figure 3.1. The upper left corner square is colored red and the colors are applied cyclically down the first column and then cyclically across each row. Now, a straight tromino will cover one square of each of the three colors wherever the tromino is placed on the board to cover three of the unit squares. On the 8×8 board, we count 21 red squares, 22 white squares, and 21 blue squares. If we are to tile the board with twenty-one straight trominoes and one monomino, it is evident that the monomino must go on one of the white squares of the tricolored board. But, will any white square do? No. Assume that we could put the one monomino on the white square in the upper right hand corner of Figure 3.1 and go on to produce the desired tiling. If we were to rotate this tiling ninety degrees counterclockwise, we would then have a tiling with the only monomino covering a red square of the tricolored board. However, we know this is impossible; we now realize that the white square cannot be one of those that covers either a red square or a blue square whenever the tricolored board is rotated or reflected onto itself. That leaves exactly the four white squares whose positions are indicated by the black squares in Figure 3.2. Check that. Will any of these four squares do? If one of the positions is good then all four are good, by symmetry. Yes, the problem can now be solved as shown in Figure 3.3.

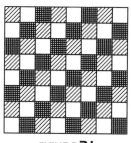

FIGURE 3.1

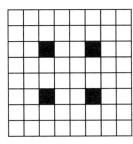

FIGURE 3.2

Sitting inside the 8×8 square in Figure 3.3 is a 5×5 square that is tiled with eight straight trominoes and one monomino. Noticing how the border was added to the 5×5 square to get the 8×8 square, we can imitate the process and expand any $n \times n$ square to an $(n+3) \times (n+3)$ square. Repeating this technique of adding a border to the 5×5 square, we see that any $(3k+2) \times (3k+2)$ square can be tiled with straight trominoes and one monomino. Starting with the 4×4 square tiled with straight trominoes and one monomino in Figure 3.3 and using the same technique, we see that any $(3k+1) \times (3k+1)$ square can be tiled with straight trominoes and one monomino. Of course, any $3k \times 3k$ square can be tiled with straight trominoes alone. We put all this together in

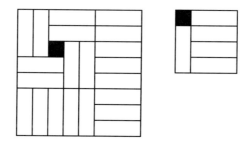

FIGURE 3.3

the following conclusion. *For any integer n with n ≥ 3, an n × n square can be tiled with straight trominoes and at most one monomino.*

Can you show that in tiling a 5 × 5 square with eight straight trominoes and one monomino you must place the monomino on the center square? The result from the previous paragraph tells us only that the following problem has a solution but does not tell what that solution is.

Problem 3.9

Where can a monomino be placed on a 7 × 7 square so that the remaining forty-eight squares can be tiled with straight trominoes?

The *L*-tromino

We now turn to the *L*-tromino, the first polyomino that is not a rectangle. It does not take long to convince yourself that the *L*-tromino does not tile a 3 × 3 square. We will want to know that the *L*-tromino does tile a 6 × 6 square. An *L*-tromino tiles a 2 × 3 rectangle; this is obvious but basic. Since two *L*-trominoes cover a 2 × 3 rectangle, any $2m \times 3n$ rectangle can be tiled easily with *L*-trominoes. The 6 × 6 square is a special case of a $2m \times 3n$ rectangle and can be tiled as suggested by Figure 3.4.

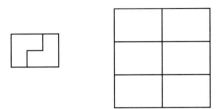

FIGURE 3.4

A 2×2 square can obviously be tiled with one L-tromino and one arbitrarily placed monomino. (In our context, the monomino must be placed on one of the four unit squares that tile the 2×2 square.) How about a 4×4 square? Yes, first place the monomino anywhere on the square. By the symmetry of rotation, it is convenient to suppose that the monomino is in the upper left 2×2 quadrant. We know we can tile the remainder of this quadrant with L-trominoes. By placing one L-tromino in the position of the black L-tromino shown in Figure 3.5, we have remaining three more quadrants, each with one square already covered. Since we can finish covering each of these with L-trominoes, we will have the 4×4 square tiled with L-trominoes and one monomino. The same procedure used for the 4×4 square now works for an 8×8 square. We will want to know this particular result for the 8×8 square later. After the monomino is arbitrarily placed, the procedure is then to place one L-tromino so that the problem is then reduced to covering with L-trominoes each of the four quadrants with one unit square deleted. That is, the procedure is to reduce the problem to four examples of the previous problem. Another application of the procedure gives a tiling of a 16×16 square with L-trominoes and one arbitrarily placed monomino. At each step we double the dimensions of the board. In general, a $2^n \times 2^n$ square can be tiled with L-trominoes and one arbitrarily placed monomino.

As dazzling as the result just proved is, the technique used to derive the result is even more dazzling. The technique is to repeatedly apply a procedure that forms new regions having some desired property by pasting together smaller regions having this property. The technique is

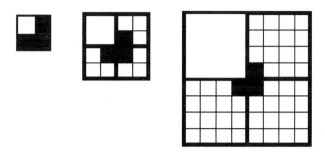

FIGURE 3.5

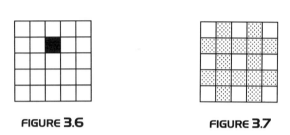

FIGURE 3.6   FIGURE 3.7

simple but powerful. You will see this technique used throughout our study of polyominoes.

It is quickly seen that a 5 × 5 square cannot be tiled with L-trominoes and one monomino when the monomino is placed in the center of the second row, as in Figure 3.6. (Covering the unit square above the monomino with an L-tromino leaves a corner square that cannot be covered.) It takes only a little longer to see that placing the monomino in either the first or second column of the second row also leads to failure. It then follows from symmetry that all the positions shaded in Figure 3.7 are prohibited if we are to have any success. Success is possible for the remaining positions. Figure 3.8 shows that success does follow if the monomino is placed on the center of the square or on the center of a side. The case that we will want to know in particular is that of placing the monomino on a corner square of the 5 × 5 square. This important case can be picked out from Figure 3.9, which shows the additional fact that a 5 × 9 rectangle can be tiled with fifteen L-trominoes. Adding six 2 × 3 rectangles, each tiled with L-trominoes, to the bottom of this 5 × 9 rectangle produces a 9 × 9 square tiled with twenty-seven L-trominoes. We will want to know this solution to Problem 3.5.

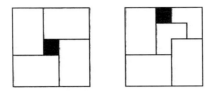

FIGURE 3.8

The tiling of the 5 × 9 rectangle in Figure 3.9 with an odd number of L-trominoes is somewhat remarkable. If you don't think so, try the very hard Problem 3.10 that is stated below. If you stretch Figure 3.9 horizontally or vertically or both ways, you will have a rectangle tiled with fifteen congruent "stretched L-trominoes." Problem 3.11 is very, very hard. It would seem the answer to the question posed there is No, but no one has been able to prove an answer to this question or to the more general question posed in Problem 3.12. Even reading Problem 3.12 takes some effort.

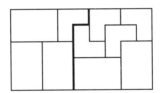

FIGURE 3.9

Problem 3.10

Can you give an example of a rectangle dissected into an odd number of congruent polygons that are neither rectangles nor stretched L-trominoes?

Problem 3.11

Is there a rectangle that can be dissected into three congruent polygons that are not rectangles?

Problem 3.12

If there is a smallest positive integer k such that k copies of a polygon tile a rectangle and if k is greater than 1, then is k necessarily even?

A *lemma* is a preliminary result proved in preparation for a theorem. We wish to prove the following lemma: *If n is a positive integer with $n \geq 6$ and an $n \times n$ square can be tiled with L-trominoes and at most one arbitrarily placed monomino, then an $(n + 6) \times (n + 6)$ square can be tiled with L-trominoes and at most one arbitrarily placed monomino.* The monomino is required for the $n \times n$ square exactly when 3 does not divide n, since squares of integers are of the form $3k$ or $3k + 1$ and since 3 divides n^2 if and only if 3 divides n. Also, since 3 divides n if and only if 3 divides $n + 6$, the monomino is needed for the $(n + 6) \times (n + 6)$ square if and only if the monomino is needed for the $n \times n$ square.

Don't be surprised if you had to read most of the paragraph above several times in order to understand what was said. Mathematics is often like that. It takes some discipline to read slowly and reread, sometimes several times. Reading mathematics is a skill that is very different from reading a newspaper. Patience and persistence are required to master the art of reading mathematics.

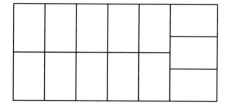

FIGURE 3.10

The proof of the lemma describes a technique for adding borders to get the new squares from the old squares. In order to prove the lemma, we first attack $6 \times n$ rectangles. Every positive integer n with $n \geq 2$ is of the form $2k$ or $2k+1$, where k is positive integer. A $6 \times 2k$ rectangle can be tiled with $2k$ copies of a 2×3 rectangle. Since $2k+1 = 2(k-1)+3$, a $6 \times (2k+1)$ rectangle can be tiled with $2k+1$ copies of a 2×3 rectangle, as in the example of the 6×13 rectangle shown in Figure 3.10. So, for $n \geq 2$, a $6 \times n$ rectangle can be tiled with $2n$ copies of an L-tromino. Therefore, we can tile the three rectangles bordering on the $n \times n$ square in Figure 3.11. When $n \geq 6$ a monomino placed on the $(n+6) \times (n+6)$ square may be supposed to be placed within the $n \times n$ square of the figure by rotating the larger square if necessary. Our desired lemma thus follows essentially from the technique of adding the borders indicated in Figure 3.11.

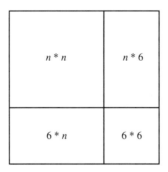

FIGURE 3.11

We are ready to prove the following theorem. *For $n \geq 2$, an $n \times n$ square can be tiled with L-trominoes and at most one monomino unless $n = 3$; if 3 does not divide n, then the position of the required monomino is arbitrary unless $n = 5$.*

Those cases where n is one of 2, 3, 4, 5, 6, 8, or 9 have already been considered. The case $n = 7$ is seen to follow from Figure 3.12, as follows. By rotation or reflection, an arbitrarily placed monomino on a

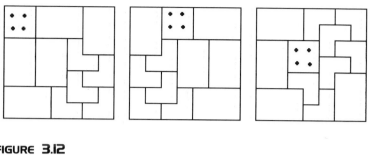

FIGURE 3.12

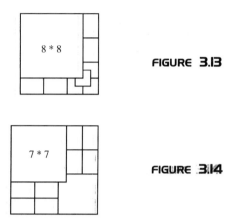

FIGURE 3.13

FIGURE 3.14

7×7 square may be supposed to cover at least one of the twelve dots in the figure. The remainder of the 2×2 square containing this dot and hence the remainder of that 7×7 square can then be covered with L-trominoes. So the theorem does hold when $n = 7$. By adding borders as indicated in the lemma, the theorem holds for all n of the form $6k + 1$, where, here and below, k is always a positive integer. Since the theorem holds for $n = 8$, then the theorem follows for all n of the form $6k + 2$ by the lemma. Since the theorem holds for $n = 9$, then the theorem follows for all n of the form $6k + 3$ by the lemma. (The case $n = 10$ does not follow from the case $n = 4$ by the lemma, since the lemma requires that $n \geq 6$.) Figure 3.13, however, shows us how to use L-trominoes to expand an 8×8 square into a 10×10 square. Since the 8×8 case was done earlier, then the theorem does hold for $n = 10$. Now, by the lemma, the theorem holds for all n of the form $6k + 4$. The case $n = 11$

is seen to follow from Figure 3.14. Note that it is a 5×5 square with a corner unit square deleted that is used in the border that expands the 7×7 square to the 11×11 square in the figure. Since the theorem holds when $n = 7$, then the theorem holds when $n = 11$. Then, by the lemma, the theorem holds for all n of the form $6k + 5$. Finally, from the trivial case $n = 6$, it follows from the lemma that the theorem holds for all n of the form $6k$. This exhausts all the possibilities for positive integers n and completes the proof of the theorem.

Introducing Reptiles

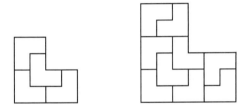

FIGURE 3.15

A polygon such that k copies tile a similar polygon is called a *reptile* of *order* k if $k > 1$. The name comes from a desire to shorten "replicating tile." Also for brevity, a reptile of order k is said to be *rep* k. Figure 3.15 shows that the L-tromino is a reptile of order 4 and also a reptile of order 9. (If all the linear dimensions of a figure are multiplied by a positive number r, then the resulting figure is said to be *similar* to the original figure with similarity ratio r; note, however, that the area of the resulting figure is r^2 times that of the original.) If an n-omino tiles a polyomino similar to itself with similarity ratio r, then the similar polygon must contain $r^2 n$ unit squares and hence must be tiled with r^2 copies of the n-omino. Since r is a rational number whose square is an integer, then r must be an integer itself. Therefore, the order of a reptile that is a polyomino is the square of an integer. (This is not true for all reptiles. For example, an isosceles right triangle is rep 2.) Problem 3.6 asks you to find the values of r for which the L-tromino is rep r^2. Using the previous theorem and Figures 3.16 and 3.17, you should be able to convince yourself that the L-tromino is rep r^2 for all positive integers r except possibly 3 and 5. Figure 3.16 applies when r is divisible by 3 and greater than 3; Figure 3.17 applies otherwise, including the cases $r = 3$ and $r = 5$ that are not covered by the theorem. Hence, we have the following answer to Problem 3.6. *The order of a reptile that is a polyomino is the square of an integer. The L-tromino is rep r^2 for each positive integer r with $r > 1$.*

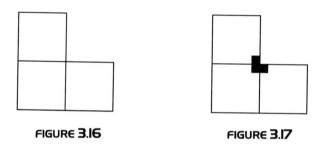

FIGURE 3.16   FIGURE 3.17

That the L-tromino is a reptile leads to a natural tiling of the first quadrant of a coordinate plane. In general, a rep k polygon is necessarily rep k^2. As k copies of a rep k reptile are put together to make a larger replica, so k copies of the larger replica are put together to make a still larger replica. There is no end to this process, which can be used to tile the first quadrant with L-trominoes. For example, Figure 3.18 shows the first three stages of forming the tiling. Each of the stages is to be viewed as sitting in the corner of the quadrant. We begin with the L-tromino in the corner of the quadrant as the first stage. Three copies are added to the first stage to form the second stage, indicated in the middle of Figure 3.18. Three copies of this second stage are added in the same manner to the second stage to form the third stage, indicated at the right in Figure 3.18. Three copies of the third stage are added in the same manner to form the fourth stage. Each stage is contained in the successive stage. So, once an L-tromino is placed that L-tromino is not moved at any later stage. The process is continued to determine a tiling of the quadrant. Figure 3.19 shows a part of this interesting tiling. The position of

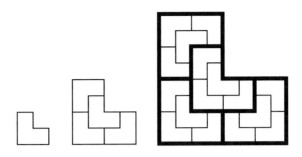

FIGURE 3.18

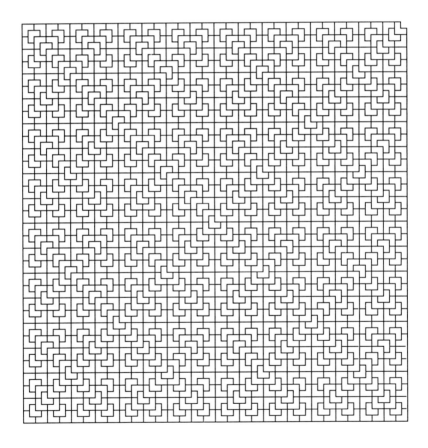

FIGURE 3.19

some of the L-trominoes in the tiling appears almost haphazard, rather than being determined by the strict rule of replicating the reptile on an increasingly larger scale. Rotating copies of the tiled quadrant about the origin, produces a tiling of the plane.

Problem 3.13

Show that a polyomino that tiles a rectangle is a reptile.

Problem 3.14

Find some more reptiles.

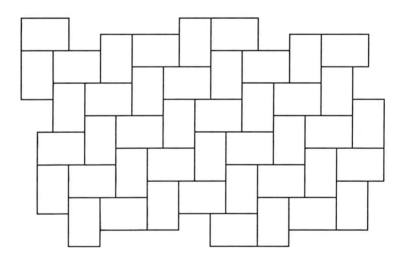

FIGURE 3.20

Figure 3.20 indicates a simple, fault-free tiling of the plane with 2×3 rectangles. Since each of the rectangles in this tiling can be dissected into two L-trominoes by sliding and rotating either one of the two tiled rectangles in Figure 3.21 onto the rectangle, you may produce an infinite number of simple, fault-free tilings of the plane with L-trominoes. You may also produce infinitely many simple, fault-free tilings of the plane with L-trominoes by pasting together copies of the infinite strips that are indicated in Figure 3.22.

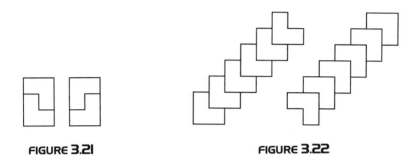

FIGURE 3.21   FIGURE 3.22

THE TETROMINOES

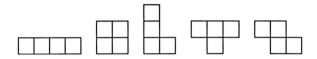

There are five tetrominoes. They are called the straight tetromino, the square tetromino, the L-tetromino, the T-tetromino, and the Z-tetromino. Some problems dealing with these polyominoes follow. Those involving a standard checkerboard are listed together, although the problems are treated in a different order in the discussion to follow. Try at least one or two that strike your fancy.

Problem 4.1

Show there is no fault-free tiling of the 8×8 checkerboard with straight tetrominoes.

Problem 4.2

Find a simple, fault-free tiling of the 8×8 checkerboard with T-tetrominoes.

Problem 4.3

Show that you cannot cover the 8×8 checkerboard with one square tetromino and fifteen T-tetrominoes.

Problem 4.4

Show that you cannot cover the 8 × 8 checkerboard with one square tetromino and fifteen *L*-tetrominoes.

Problem 4.5

Show that you cannot cover the 8 × 8 checkerboard with one square tetromino and fifteen other tetrominoes selected from the straight tetrominoes and the *Z*-tetrominoes.

Problem 4.6

Can you tile a 10 × 10 square with straight tetrominoes?

Problem 4.7

Can you tile a 10 × 10 square with *L*-tetrominoes?

Problem 4.8

Can you tile a 10 × 10 square with *T*-tetrominoes?

Problem 4.9 ·

Which tetrominoes are reptiles?

Two of the five tetrominoes are rectangles. The square tetromino tiles the standard 8 × 8 checkerboard exactly one way, which is shown in Figure 4.1.

The straight tetromino tiles the checkerboard in several ways. The obvious way is shown in Figure 4.2, where all the straight tetrominoes have the same aspect. This tiling certainly has a fault line. It is not quite so obvious that any tiling of the checkerboard with straight tetrominoes has a fault line. However, it is true and the analysis of the problem is easy. In Figure 4.3 the unit square marked 1 can be covered in just two ways, which are equivalent initial positions by the symmetry of a reflection in a diagonal. Then the square marked 2 in the figure can be covered in only one way without forming a fault line. The same goes for the squares marked 3 through 9, and this ninth straight tetromino introduces a fault line no matter how the tiling is completed.

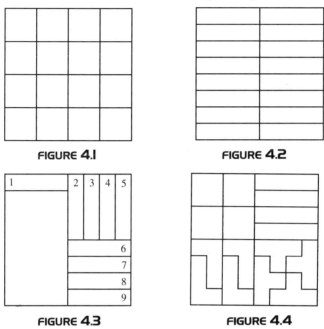

FIGURE 4.1   FIGURE 4.2

FIGURE 4.3   FIGURE 4.4

The rectangular tetrominoes are not the only tetrominoes to tile the checkerboard. In fact, four of the five tetrominoes tile a 4×4 square, as shown in Figure 4.4, and hence each of these tetrominoes tiles the checkerboard.

Rectangles Tiling Rectangles

In addition to the standard checkerboard, which rectangles does a straight tetromino tile? We will look at the more general problem of which $n \times 1$ rectangles tile an $h \times w$ rectangle. If n divides h or if n divides w, then the straight n-ominoes, which are the $n \times 1$ rectangles, tile an $h \times w$ rectangle in an obvious way, with all the n-ominoes having the same aspect. We will prove the converse below. The result will then say that if a rectangle does not have an obvious tiling with straight n-ominoes

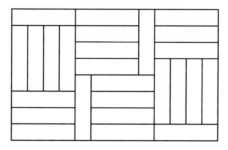

FIGURE **4.5**

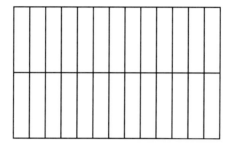

FIGURE **4.6**

then the rectangle does not have any tiling with straight n-ominoes. The tiling of the rectangle in Figure 4.5 thus implies the existence of the less exciting tiling of the same rectangle in Figure 4.6.

The converse that we need to show states: If a straight n-omino tiles an $h \times w$ rectangle, then n divides h or w. Note that "n divides h or w" is a stronger requirement than "n divides hw." For example, since 4 does not divide 10, then a 10×10 square cannot be tiled with straight tetrominoes, even though 4 does divide 100. The proof of the converse is a little more sophisticated than most of the arguments we will see. Yet the basic idea is fairly simple. We can often deduce information by calculating something in two different ways and setting these values equal to each other. You may find it informative to follow the general argument given below with the specific example mentioned above in mind. That is, assume a 10×10 rectangle is tilled with twenty-five copies of the straight 4-omino and follow the proof through to reach the contradiction that 4 divides 10, thus showing that the assumption is necessarily false.

To begin our general proof, suppose there is an $h \times w$ rectangle that is tiled with straight n-ominoes and that n does not divide w. We will show that n must then divide h. Since n does not divide w, then, in attempting to divide n into w, we get a quotient q and a nonzero remainder r that is less than n. So $w = nq+r$ where $0 < r < n$. We color the w columns of the $h \times w$ rectangle cyclically with n distinct colors. All of the h unit squares in one column have the same color. The $h \times w$ rectangle is dissected into q copies of an $h \times n$ rectangle having exactly one column of each of the n colors and one copy of an $h \times r$ rectangle having one column of each of the first r colors only. In particular, the nth color does not occur in the $h \times r$ rectangle. (In the example suggested above where $h = w = 10$ and $n = 4$ (and so $r = q = 2$), we now have the ten columns colored: red, white, blue, green; red, white, blue, green; red, white. There are three red columns but only two green columns. See Figure 4.7 for a similar example where $w = 14$, $n = 4$, $q = 3$, and $r = 2$.) Let c_i be the number of unit squares of the $h \times w$ rectangle colored with the ith color. Then $c_1 = h(q+1)$ and $c_n = hq$, since there are $q + 1$ columns of color number 1 and there are q columns of color number n and since there are h unit squares in each column. Next, we use one of the great techniques in mathematics of counting the same thing in two different ways and then setting the calculations equal to each other. For the second way of calculating c_1 and c_n, we note that a straight n-omino placed on the colored $h \times w$ rectangle will either cover exactly one square of each of the n colors when placed horizontally or else will cover n squares of the same color when placed vertically. Let x be the number of n-ominoes placed horizontally in the tiling, and let y_i be the number placed vertically on a column colored with the ith color. Then $c_1 = x + ny_1$ and $c_n = x + ny_n$. Equating the two values of $c_1 - c_n$ that we get from the two ways of calculating each of c_1 and c_n, we have

$$h = [h(q + 1)] - [hq] = c_1 - c_n$$
$$= [x + ny_1] - [x + ny_n]$$
$$= n(y_1 - y_n).$$

Since n divides the right-hand side, then n must divide the left-hand side. So n divides h, as desired. This completes the proof of the following

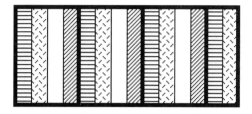

FIGURE 4.7

theorem: *A straight n-omino tiles an h × w rectangle if and only if n divides h or w.*

Once you have learned that a certain tiling problem has a solution, the question of how many solutions exist rears its ugly head. This is often such a difficult question that the ugly head is simply ignored. However, the problem of finding the number of tilings of the standard checkerboard with straight tetrominoes is within reach. The two facts that you need to know are that such tilings always have a fault line and that the straight tetromino tiles an $h \times w$ rectangle if and only if 4 divides h or w. Don't forget to take symmetry into consideration—and this is the hard part—if you attack Problem 4.10. Problems 4.10, 4.11, and 4.12 are increasingly difficult and are not solved here. Two tilings in these problems are counted as the same "way" if one can be obtained from the other by a rotation or a reflection. In general, two tilings are considered equivalent if they are congruent.

Problem 4.10

In how many ways can the 8×8 checkerboard be tiled with straight tetrominoes?

Problem 4.11

In how many ways can the 8×8 checkerboard be tiled using both straight tetrominoes and square tetrominoes?

Problem 4.12

In how many ways can the 8×8 checkerboard be tiled with rectangular polyominoes?

The three problems above suggest the following three unsolved problems. We will make only a dent in the second of these by finding conditions that determine whether a particular rectangular polyomino tiles a given rectangle.

Problem 4.13

In how many ways can an $n \times n$ square be tiled with square polyominoes?

Problem 4.14

In how many ways can an $h \times w$ rectangle be tiled with an $a \times b$ rectangular polyomino?

Problem 4.15

In how many ways can an $h \times w$ rectangle be tiled with rectangular polyominoes?

A *corollary* is a result that follows immediately from a theorem. We will use the theorem concerning which straight n-ominoes tile a given rectangle to answer the more general question of which $a \times b$ rectangles tile a given $h \times w$ rectangle. Of course, you could take the attitude that the theorem above is really a lemma that is used to prove the more important result that we are going to call a corollary. Both results could be called theorems, too. You see, all this name calling is a matter of opinion. Whatever the nomenclature, the result to be proved is the following. *An $a \times b$ rectangle tiles an $h \times w$ rectangle if and only if a divides h or w, b divides h or w, and each of h and w can be expressed in the form $xa + yb$ where x and y are nonnegative integers.*

Suppose the $h \times w$ rectangle is tiled with the $a \times b$ rectangles. We first note that since each $a \times b$ rectangle can be dissected into b of the

$a \times 1$ rectangles, then a must divide h or w by the theorem. Similarly, since each $a \times b$ rectangle can be dissected into a of the $b \times 1$ rectangles, then b must divide h or w by the theorem. Finally, each of h and w must be of the form $xa + yb$, since an edge of the larger rectangle is covered by the edges of the smaller rectangles. This proves the "only if" part of the corollary.

We now prove the converse. We suppose that a divides h or w, that b divides h or w, and that each of h and w is of the form $xa + yb$ with x and y nonnegative integers. If a divides one of h or w and if b divides the other, then the $h \times w$ rectangle can be tiled with the $a \times b$ rectangles in the obvious way, where all the smaller rectangles have the same aspect. Now suppose both a and b divide h. Since $w = xa + yb$, then the $h \times w$ rectangle can be covered with x copies of an $h \times a$ rectangle and y copies of an $h \times b$ rectangle. These rectangles can then easily be dissected into $a \times b$ rectangles to form the desired tiling. Likewise, if both a and b divide w with $h = xa + yb$, then the $h \times w$ rectangle can be covered with x copies of a $w \times a$ rectangle and y copies of a $w \times b$ rectangle. These rectangles can then easily be dissected into $a \times b$ rectangles to form the desired tiling. This finishes the proof of the corollary. We note that the proof tells us that if there is any tiling, then there is a tiling where all the $a \times b$ rectangles having the same aspect together form a rectangle.

The L-tetromino

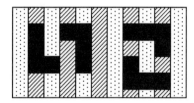

FIGURE **4.8**

What rectangles can be tiled with the L-tetromino? Suppose we have an $h \times w$ rectangle tiled with L-tetrominoes. So h and w are both greater than 1. Since 4 divides hw, at least one of h or w must be even. Without loss of generality, we may suppose w is even. We color the columns of the $h \times w$ rectangle cyclically with two distinct colors, say orange and green. All of the h unit squares in one column have the same color. Now, an L-tetromino covers three unit squares of one color and one unit square of the other color wherever the tetromino is placed on the rectangle, as is illustrated in Figure 4.8. Let x be the number of L-tetrominoes that cover three orange squares and one green square. Let y be the number of L-tetrominoes that cover one orange square and three green squares. So $x + y = (hw)/4$. The number of unit squares on the $h \times w$ rectangle that are orange equals the number that are green. That is, $3x + y = x + 3y$. So, $x = y$. Therefore, the number of tiles in the tiling is $2x = x + y = hw/4$. From this last equation, we see that $8x = hw$. We conclude that 8 divides hw.

FIGURE 4.9

We now show that, conversely, if h and w are both greater than 1 and if 8 divides hw, then an $h \times w$ rectangle can be tiled with L-tetrominoes. Is it clear that the $h \times w$ rectangle must be either an $8m \times n$ rectangle with n odd or else a $2m \times 4n$ rectangle? (There are only two ways to put three indistinguishable eggs into two indistinguishable baskets: put all three eggs in one basket or else put one egg in one basket and two eggs in the other. In case you are having a hard time figuring out what eggs have to do with anything, the "baskets" are the sides of the rectangle and the "eggs" are the three 2's that are the factors of 8.) Since two L-tetrominoes tile a 2×4 rectangle, the $2m \times 4n$ rectangles are no problem because they can easily be tiled with mn copies of the 2×4 rectangle with all copies having the same aspect. (In particular, the $8m \times 2r$ rectangles are included among these.) The tiling of the 2×4 rectangle with L-tetrominoes can be extended to a tiling of the 8×3 rectangle with L-tetrominoes as is shown in Figure 4.9. The remaining case of an $8m \times (2k + 3)$ rectangle, where k is a nonnegative integer, follows from the $8 \times (2k+3)$ case illustrated by Figure 4.10. This finishes the proof of the following theorem. *An $h \times w$ rectangle can be tiled with L-tetrominoes if and only if*

1. *each of h and w is greater than 1 and*
2. *8 divides hw.*

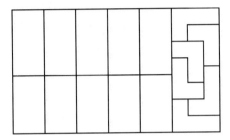

FIGURE 4.10

The T-tetromino

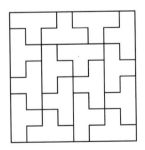

FIGURE 4.11

A fault-free tiling of the 8×8 checkerboard with T-tetrominoes is given in Figure 4.11. Suppose we have a tiling of an $h \times w$ rectangle with T-tetrominoes. Then, since 4 must divide hw, at least one of h or w must be even. Therefore, in the ordinary checkerboard coloring of the rectangle, the number of black squares equals the number of white squares. (We are still a little ill over the green and orange squares and have returned to the standard colors.) Now, wherever a T-tetromino is placed on the checkerboard, the T-tetromino will cover three unit squares of one color and one unit square of the other color. Let x be the number of T-tetrominoes that cover three black squares and one white square. Let y be the number of T-tetrominoes that cover one black square and three white squares. So $x + y = (hw)/4$. Since the number of unit squares on the $h \times w$ rectangle that are black equals the number that are white, then $3x + y = x + 3y$. So, $x = y$. Therefore, the number of tiles in the tiling is $2x = x + y = hw/4$. From this last equation, we see that $8x = hw$. We conclude that 8 divides hw and that the number of T-tetrominoes in the tiling is even. In particular, a 10×10 square cannot be tiled with T-tetrominoes.

You have undoubtedly noticed the similarity between the proof in the previous paragraph and the corresponding result for the L-tetrominoes. However, there we were able to prove the converse; here we are not. In fact, it is impossible to tile a 10×8 rectangle with T-tetrominoes, even though 8 divides 80. The theorem that applies but which we are not going to prove is the following. *An $h \times w$ rectangle can be tiled with T-tetrominoes if and only if 4 divides each of h and w.*

If each of h and w are integral multiples of 4, as is required in the theorem just stated, then there is a trivial tiling of the rectangle with T-tetrominoes, since the T-tetromino tiles a 4×4 square. As so often happens with the tetrominoes, there is either an easy solution to a tiling problem or there is no solution at all. D. W. Walkup's proof of the theorem stated above can be found on pages 986–988 of volume 72 (November 1965) of *The American Mathematical Monthly.* This journal is published by The Mathematical Association of America and can be found in most college libraries. The desired proof is two and one-half pages long. The proof is more complicated than those proofs that are presented here but is within reach of anyone who has come this far and has the desire to search out the proof. You are encouraged to make the

effort. You will know you have succeeded if you can present the argument to a friend and convince your friend of the result. If you do this,
give yourself three giant points.

The Z-tetromino

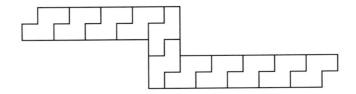

FIGURE 4.12

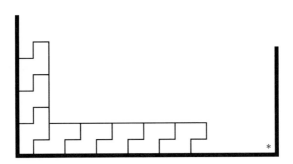

FIGURE 4.13

The Z-tetromino is the first of our polyominoes not to tile a rectangle. It is true that the Z-tetromino can get around alternate right and
left turns, as is suggested by Figure 4.12. However, a Z-tetromino cannot make two successive turns in the same direction. To see this consider Figure 4.13. The Z-tetromino that covers the corner in the figure
is in one of the two equivalent possible positions. The neighboring Z-

tetrominoes along the edges are determined, as are the neighbors of neighbors. Everything is fine until the Z's come up to an adjacent corner of the rectangle, where there is disaster. If we can't get two adjacent corners covered, we will never get the rectangle covered. *The Z-tetromino does not tile any rectangle.* On the other hand, copies of the "infinite Z-strip" and the "infinite L-strip" that are suggested by Figures 4.12 and 4.13 can be pasted together to form tilings of the plane. Can you form a tiling of the plane with the Z-tetromino that contains each of these two infinite strips?

There are many problems that you can make up by asking for tilings of the checkerboard with various combinations of the tetrominoes. The impossibility of some of these can be shown by clever colorings of the checkerboard. With the ordinary coloring of the checkerboard, a T-tetromino covers three squares of one color and one of the other color, while a square tetromino covers two of each color. Suppose we could tile the checkerboard with one square tetromino and fifteen T-tetrominoes. Let x be the number of T-tetrominoes that cover three black squares and one white square. Let y be the number of T-tetrominoes that cover one black square and three white squares. This time we have $3x+y+2 = x+3y+2$ and $x+y = 15$, which is impossible since x and y are integers. Replacing the square tetromino in Problem 4.3 by the Z-tetromino or by the straight tetromino also leads to an impossibility by the same reasoning. Under the alternate coloring of the columns with two colors, the L-tetromino covers three of one color and one of the other while the square tetromino covers two of each color. The same equations as those above now tell us that the checkerboard cannot be tiled with fifteen L-tetrominoes and one square tetromino. Likewise, fifteen L-tetrominoes and one Z-tetromino will not work. Problem 4.5 asks us to show that one square tetromino and fifteen other tetrominoes selected from the straight tetrominoes and the Z-tetrominoes is another impossibility. Here the clever coloring is shown in Figure 4.14. The result follows since the straight tetromino and the Z-tetromino always cover an even number of squares of either color, while the square covers three of one and one of the other. There is no limit to the problems that you can compose if you allow larger square boards or even rectangular boards. Of course, you will probably be able to ask questions that you will not be able to answer.

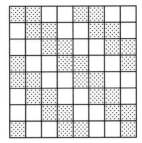

FIGURE 4.14

More on Reptiles

Each of the five tetrominoes tiles an infinite strip of width 2. See Figure 4.15. For each tetromino, copies of this strip can be pasted together to form infinitely many tilings of the plane.

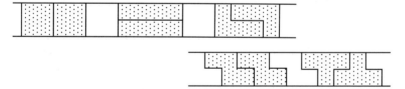

FIGURE 4.15

The Z-tetromino is not a reptile because this tetromino cannot cover two adjacent corners of a rectangle. The Z-tetromino is the first of our polyominoes not to be a reptile. The other four tetrominoes are reptiles since each tiles a square. That these four tetrominoes are reptiles follows from the argument in the next paragraph. Figure 4.16 shows the tiling for the T-tetromino reptile.

A polyomino that tiles a rectangle is a reptile. To see this, first observe that if an n-omino P tiles an $h \times w$ rectangle then hw copies of this rectangle tile a square with side hw. Hence, P tiles the square. It should be remarked that this may not be the smallest square that is tiled by the rectangle, as in the case $w = 2h > 2$. In any case, a polyomino that

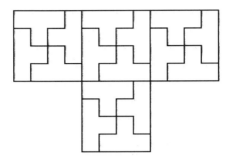

FIGURE **4.16**

tiles a rectangle also tiles a square. Now, n copies of this square form a polyomino Q that is similar to the original n-omino P. Since P tiles the square, then P tiles Q, as desired.

If you care to keep track of the numbers in the proof above, the argument shows that P is rep $(hw)^2$. Of course, the given polyomino may be rep k for smaller values of k as well. The L-tetromino is an example and is rep 4. The straight tetromino and the square tetromino are rectangles themselves and each is obviously rep 4. Figure 4.17 shows the three rep 4 tetrominoes. You may want to convince yourself that the T-tetromino is not rep 9. Which of the tetrominoes are rep 9?

FIGURE **4.17**

Somewhat along the same line as the argument above, you should convince yourself of the following. *If a polygon is rep m and rep n, then the polygon is rep mn.* Would you expect that taking m copies of polygon P to form polygon Q and then n copies of Q tiled with P to form polygon R produces the same tiling of R with P as does starting with n copies

of P to form polygon S and then m copies of S tiled with P to form R?
You may have to read the last sentence several times for its meaning to
sink in. However, once the meaning is clear the answer is a clear No.
If all else fails, try the constructions with an example. The L-tetromino
can be used for an example but the L-tromino, which is rep 4 and rep 9,
provides an easier example.

The tiling of a 4×4 square with the T-tetromino gives a clue to a
general method for producing reptiles. The method actually produces
polyominoes such that four copies tile a square. From the center of a
$2t \times 2t$ square polyomino, trace a path along the edges of the unit squares
to an edge of the $2t \times 2t$ square in such a way that the path does not intersect itself or, except for the center, any of its images under rotations
about the center of 90°, 180°, or 270°. The path and these three images form the tiling of the square with four copies of a polyomino. The
polyomino is a t^2-omino that is rep $4t^2$. Figure 4.18 shows an example
with $t = 4$. One of the ugly heads mentioned earlier rears to ask how
many such t^2-ominoes are there for a given t or even how many of the
described paths are there for a given value of t. Once more, the ugly
head is ignored.

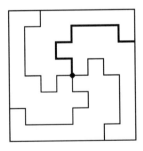

FIGURE **4.18**

THE PENTOMINOES

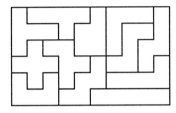

As Goldilocks might say, "The tetrominoes are too few, the hexominoes are too many, but the number of pentominoes is just right." The dozen pentominoes are the well from which runs a seemingly endless stream of puzzles. It is this wealth of pentomino puzzles that has attracted the layman and the professional mathematician alike to polyominoes. We will state only thirty such puzzles that involve all twelve pentominoes. No new mathematical concepts are introduced in this chapter; the entire chapter is devoted to puzzles and solutions. Many of these puzzles first appeared from the 1920s through the 1950s in the journal *The Fairy Chess Review*, where it is stated, "The possibilities of the twelve fives are not infinite but they will provide years of amusement." The time reference is not exaggerated; like jigsaw puzzles, these puzzles are for the most part as pleasing to solve the second time around as they were when new. Like all good puzzles they are not solved quickly.

It is only proper to begin with what has been called the oldest of the pentomino puzzles. This is Puzzle 74 from Dudeney's book *The Canter-*

bury Puzzles of 1907 and our Problem 5.1. The puzzle comes with a story. The son of William the Conqueror was winning against the dauphin of France at chess to the extent that the dauphin threw the chessmen at his opponent, who then broke the chessboard over the dauphin's head. In this turmoil, the chessboard broke into the thirteen pieces shown in Figure 5.1. The pieces are the twelve pentominoes and the square tetromino, colored as they might be cut from a chessboard. In addition to solving the problem of putting the board together again, we will also solve the more general Problem 5.2, where as usual we ignore the coloring of the pentominoes of Figure 5.1 and allow the pieces to be flipped over.

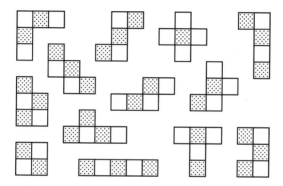

FIGURE **5.1**

Problem 5.1

Put the thirteen pieces in Figure 5.1 together to form the standard chessboard.

Problem 5.2

Form an 8×8 square with the twelve pentominoes and the square tetromino. Show that the square tetromino may be placed anywhere within the 8×8 square.

Problem 5.3

Form an 8×8 square with the twelve pentominoes and the straight tetromino. Show that the straight tetromino may be placed anywhere within the 8×8 square.

Problem 5.4

In forming an 8×8 square with the twelve pentominoes and two dominoes, where can the dominoes be placed within the 8×8 square?

The twelve pentominoes together cover sixty unit squares; the four additional unit squares needed to cover a checkerboard could be any one of the tetrominoes. Problems 5.2 and 5.3 are examples. Analogous problems deal with the L-tetromino, the T-tetromino, and the Z-tetromino; however, here there are unsolvable positions for the tetromino. Also, think of the number of puzzles that you can make up by arbitrarily placing four monominoes on the board. In many of these puzzles, the four monominoes are placed to form a symmetrical pattern. In looking at Problem 5.4, note that the dominoes may isolate a set of unit squares only if the number of these unit squares is divisible by 5. However, the number of possibilities in Problem 5.4 is so large that the problem will probably be left to a computer.

Rectangle Puzzles

The most common introduction to polyominoes is one of the puzzles that ask you to form a rectangle from the twelve pentominoes. These puzzles are the most basic and the best. From the arithmetic, there are several possibilities. Although forming a 1×60 rectangle or a 2×30 rectangle with the pentominoes is out of the question, the 3×20, the 4×15, the 5×12, and the 6×10 rectangles are not only arithmetic possibilities but provide solvable puzzles. There are 2339 solutions for the 6×10 rectangle. There are 1010 solutions for the 5×12 rectangle. There are 368 solutions for the 4×15 rectangle. There are 2 solutions for the 3×20 rectangle. Since there are so many solutions for the 6×10

rectangle, some additional restrictions may be imposed, such as those in Problems 5.9, 5.10, and 5.11. All thirty-two possible cases for Problem 5.11 have solutions.

Problem 5.5

Use the twelve pentominoes to form a 6×10 rectangle.

Problem 5.6

Use the twelve pentominoes to form a 5×12 rectangle.

Problem 5.7

Use the twelve pentominoes to form a 4×15 rectangle.

Problem 5.8

Use the twelve pentominoes to form a 3×20 rectangle.

Problem 5.9

Use the twelve pentominoes to form a 6×10 rectangle with a 4×5 subrectangle.

Problem 5.10

Use the twelve pentominoes to form a 6×10 rectangle, where each of the pieces touches the border of the rectangle.

Problem 5.11

Throw the twelve pentominoes onto a table at random. Then form a 6×10 rectangle without turning over any of the pentominoes.

Another set of simply stated but difficult rectangle problems has to do with building two rectangles at the same time with the twelve pentominoes. If the rectangle problems are the best of the problems using all the pentominoes, then the best of the best are those that ask for simultaneous rectangles. Except for Problem 5.16, which has 112 solutions, these problems have a small number of solutions. Some of these might even take days to do. But then, that is the challenge.

Problem 5.12

Use the twelve pentominoes to form simultaneously a 3×5 rectangle and a 5×9 rectangle.

Problem 5.13

Use the twelve pentominoes to form simultaneously a 4×5 rectangle and a 4×10 rectangle.

Problem 5.14

Use the twelve pentominoes to form simultaneously a 5×5 rectangle and a 5×7 rectangle.

Problem 5.15

Use the twelve pentominoes to form simultaneously two 5×6 rectangles.

Problem 5.16

Use the twelve pentominoes to form simultaneously a 1×5 rectangle and a 5×11 rectangle.

Problem 5.17

Show that the twelve pentominoes cannot be used to form simultaneously three 4×5 rectangles.

Problem 5.18

Use the twelve pentominoes and three monominoes to form simultaneously three 3 × 7 rectangles.

Using the twelve pentominoes to form simultaneously a 2 × 10 rectangle and either a 4 × 10 rectangle or a 5 × 8 rectangle is impossible. Likewise, the combination of a 3 × 10 rectangle and a 5 × 6 rectangle and the combination of a 4 × 5 rectangle and a 5 × 8 rectangle are each impossible.

Other Pentomino Puzzles

Many regions of sixty unit squares have been proposed for pentomino puzzles. Some, such as those shown in Figures 5.2 and 5.3, can be solved. Some, such as those shown in Figures 5.4, 5.5, and 5.6, are impossible to solve. For some, such as that in Figure 5.7, it is unknown whether there is a solution or not; this particular region was given by Golomb in his book as Problem 81.

Problem 5.19

For each of the two patterns shown in Figure 5.2, use the twelve pentominoes to form that pattern.

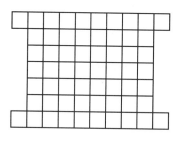
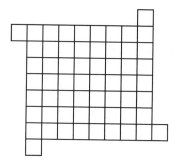

FIGURE 5.2

Problem 5.20

Use the twelve pentominoes to form the pattern shown in Figure 5.3.

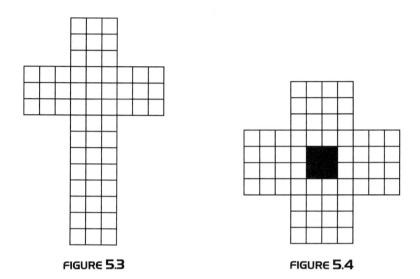

FIGURE 5.3   FIGURE 5.4

Problem 5.21

Show that the twelve pentominoes cannot form the pattern shown in Figure 5.5.

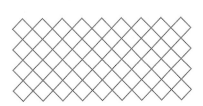

FIGURE 5.5

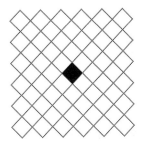

FIGURE 5.6

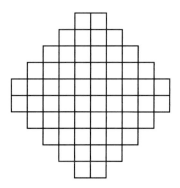

FIGURE 5.7

Problem 5.22

Can the twelve pentominoes form the pattern shown in Figure 5.7?

Problem 5.23

Use the twelve pentominoes to form a pattern that can be folded to cover the surface of a cube.

The Double Duplication Problem asks that two pentominoes form the same shape as two other pentominoes and that the remaining eight pentominoes form a shape that is similar to this shape. One solution is shown in Figure 5.8. Three more solutions come from the same figure by flipping one or both of the FT region and the PY region. Regions will often be denoted by the pentominoes that fill them. For example, the FT region in Figure 5.8 is the union of the F-pentomino and the T-pentomino. If you cannot name each of the twelve pentominoes, look back at Figure 1.9 and recall $FLIP$ ’N $TUVWXYZ$.

Problem 5.24

Find another solution to the Double Duplication Problem.

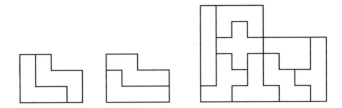

FIGURE **5.8**

The Triplication Problem asks that, given a pentomino, nine of the remaining pentominoes be used to form a figure that is similar to the given pentomino. A solution for the Triplication Problem when given the X-pentomino can be seen in Figure 5.9.

FIGURE **5.9**

Problem 5.25

For each pentomino, solve the Triplication Problem.

The Ten Problem or the Three Congruent Pairs Problem asks that the twelve pentominoes be divided into three sets of four and that each of these sets be divided into two pairs in such a way that each pair covers the same region of ten unit squares. One solution to the Ten Problem is shown in Figure 5.10.

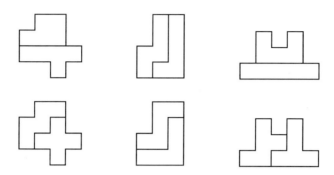

FIGURE 5.IO

Problem 5.26

Find another solution to the Ten Problem.

The Fifteen Problem asks that the twelve pentominoes be divided into four sets of three such that each set covers the same region of fifteen unit squares. This is Golomb's problem 88. No solution is known to the Fifteen Problem; nor has anyone proved the problem is impossible.

Problem 5.27

Be the first to find a solution to the Fifteen Problem or to show there is no solution.

The Twenty Problem asks that the twelve pentominoes be divided into three sets of four such that each set covers the same region of twenty unit squares. One solution to the Twenty Problem is shown in Figure 5.11.

Problem 5.28

Find another solution to the Twenty Problem.

The Fifteen–Sixty Problem asks for a figure covered by the twelve pentominoes and which contains a similar figure composed of three

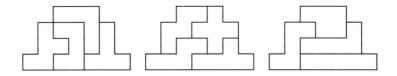

FIGURE 5.11

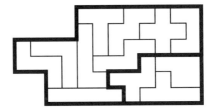

FIGURE 5.12

pentominoes. There are fifty-six solutions where the three pentominoes form a 3×5 subrectangle of a 6×10 rectangle. Another solution is given in Figure 5.12.

Problem 5.29

Find another solution to the Fifteen–Sixty Problem.

Problem 5.30

What is the least number of pentominoes that span the checkerboard? That is, what is the least number of pentominoes that can be placed on an 8×8 square such that none of the remaining pentominoes can be added to the board?

Pentomino Games

The last problem suggests an interesting game. The requirements are a standard checkerboard and a set of twelve pentominoes of the appropriate size. Taking turns, each player places one pentomino on the board until no play is possible. The loser is the person who cannot play, either because there is no space for any of the remaining pentominoes or because there are no more pentominoes. The games are short and there are no draws. There are at least five moves and at most twelve moves. The I, L, U, V, and Y can be placed on the checkerboard so that no other pentomino will fit. A variation of the game is for each player to choose six pentominoes before playing, with the last to choose being the one to play first. Another variation requires that the pentominoes be dealt at random to the players before play begins.

A different game that uses pentominoes is a generalization of tic-tac-toe. Instead of aiming for "three in a row," as you do in tic-tac-toe, you aim for the pentomino that is the agreed upon goal. The X-player and O-player take turns marking the unit squares of a checkerboard until one or the other completes forming the pentomino with all X's or all O's. The problem with this game is the same as the problem with tic-tac-toe. There are strategies that allow a knowing player if not to win, then at least never to lose. If you are interested, look up Martin Gardner's column in the April 1979 *Scientific American.* You might consider complicating the game by allowing the players to choose their own pentominoes for a goal; you have to decide if the goals are to be announced before play begins or are to be secretly recorded first and then revealed only after a declaration of winning.

Some Solutions

Let's return to some of the problems stated above. Figure 5.13 gives Dudeney's solution to the broken chessboard, our Problem 5.1.

In covering the checkerboard with the twelve pentominoes and the square tetromino, there are, up to symmetry, ten places to place the tetromino. To see this, consider Figure 5.14. The center of the tetromino can be any one of the forty-nine positions where four unit squares come together on the board. However, by a rotation about the center of the checkerboard, we may suppose the center of the tetromino is in the upper left-hand corner of the board. Further, by a reflection in the main diagonal we may suppose the center is on or above the main diagonal. So, if we can solve the problem for each of the ten cases where the center of the square tetromino is one of the ten dots in Figure 5.14,

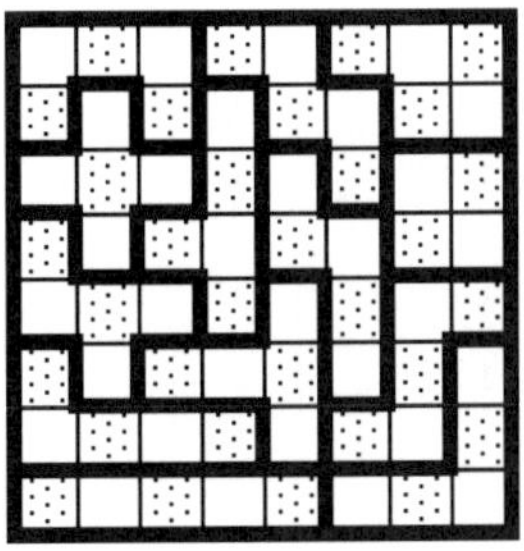

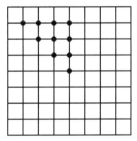

FIGURE 5.13

FIGURE 5.14

then we can solve the problem for any position of the tetromino on the board. The ten cases can be further reduced by considering Figures 5.15 and 5.16. The square tetromino and the V-pentomino can be combined to form a 3×3 square, which can be rotated to any one of the four relative positions shown in Figure 5.15. Figure 5.16 shows three 8×8 squares, each covered by a 3×3 square and the set of eleven pentominoes that excludes the V-pentomino. By matching the dots in Figures 5.15 and 5.16, we have a solution for each of the ten cases that are necessary to show the square tetromino can be arbitrarily placed on the checkerboard, with the pentominoes filling the remainder of the board.

Figure 5.17 shows the only two solutions to Problem 5.10, where each pentomino touches the border of the 6×10 rectangle. Figure 5.18 gives a 4×15 rectangle. Figure 5.19 gives one of the two solutions to forming a 3×20 rectangle. The other solution to Problem 5.8 is obtained by rotating the $LNFTWYZ$ region of Figure 5.19.

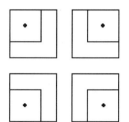

FIGURE 5.15

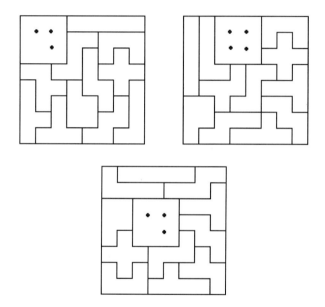

FIGURE 5.16

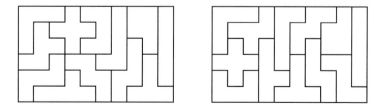

FIGURE 5.17

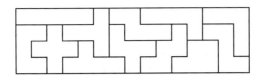

FIGURE 5.18

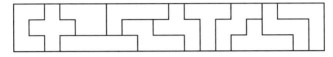

FIGURE 5.19

We now come to those problems that require forming two rectangles at the same time with all the pentominoes. Figure 5.20 shows the unique solution to using the twelve pentominoes to form simultaneously a 5 × 3 rectangle and a 5 × 9 rectangle.

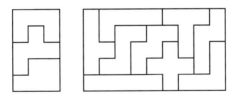

FIGURE 5.20

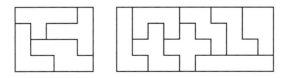

FIGURE 5.21

Figure 5.21 shows one of the five solutions to using the twelve pentominoes to form simultaneously a 4×5 rectangle and a 4×10 rectangle. The other four solutions are found in turn as follows. In Figure 5.21, flip the FX region to get the second solution. In the second solution, flip the LUX region to get the third. In the third solution, flip the FX region to get the fourth. In the fourth solution, interchange the PZ region and the FU region to get the fifth and final solution.

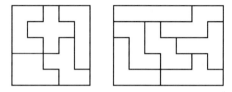

FIGURE 5.22

Figure 5.22 shows the unique solution to using the twelve pentominoes to form simultaneously a 5×5 rectangle and a 5×7 rectangle.

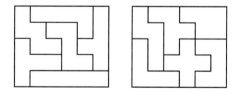

FIGURE 5.23

Figure 5.23 shows one of the two solutions to forming simultaneously 5×6 rectangles. The only other solution is obtained by rearranging the F-pentomino and the N-pentomino within the FN region in Figure 5.23.

Assume the twelve pentominoes simultaneously cover three 4×5 rectangles. The X-pentomino must be in one of these rectangles. Fur-

FIGURE 5.24

ther, the center of the X-pentomino must lie in one of the six unit squares in the interior of the 4×5 rectangle. By symmetry, much like the argument for Problem 5.2, we see that only the two cases indicated by the dots in Figure 5.24 need be considered. The first case is immediately discarded because the X-pentomino isolates sets of one and fourteen unit squares, neither of which can be covered by pentominoes. The position of the X-pentomino in the second case in Figure 5.24 requires the U-pentomino on one side and the L-pentomino on the other. By symmetry, it makes no difference which sides are covered by the U-pentomino and the L-pentomino. However, what is then left uncovered is a union of five squares that is congruent to the already used L-pentomino. Therefore, since we cannot form one 4×5 rectangle containing the X-pentomino, then we cannot simultaneously cover three 4×5 rectangles using the twelve pentominoes.

Figure 5.25 shows one of the two solutions to forming simultaneously three 3×7 rectangles with the twelve pentominoes and three monominoes. The other solution to Problem 5.18 is to flip the region covered by the Y-pentomino and the adjacent monomino in Figure 5.25.

Figure 5.26 solves Problem 5.20. That Figures 5.4 and 5.6 give regions that cannot be covered with the pentominoes can be shown by the tedious method of backtracking. We will look into this method in the next chapter. For now, we will be happy with the clever argument that shows the region in Figure 5.5 cannot be covered with the twelve pentominoes. Look back at the figure and count the squares that are on the border. You should get 22. Count the maximum number of such border squares that each pentomino could cover. ($FLIP\ N\ TUVWXYZ$ ac-

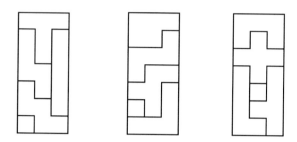

FIGURE 5.25

counts for 3, 1, 1, 2; 2; 1, 1, 1, 3, 3, 2, 1, respectively.) Now, adding these numbers, you should get 21. Close, but no cigar.

The singular problem on pentominoes from *The Fairy Chess Review* is our Problem 5.23, Use the twelve pentominoes to form a pattern that can be folded to cover the surface of a cube. Note that Figure 5.26 does not lead to a solution. This fiendishly difficult problem is solved in Figure 5.27.

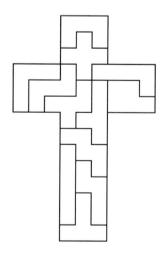

FIGURE 5.26

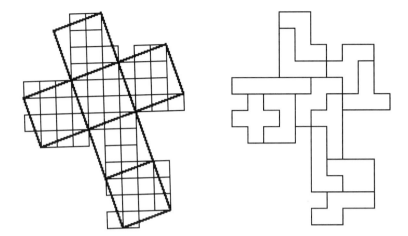

FIGURE 5.27

THE POLYPOIC PENTOMINO

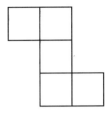

The I-pentomino is a rectangle. Two copies of either the L-pentomino or the P-pentomino tile a 2×5 rectangle. See Figure 6.1. Of the remaining nine pentominoes, it is fairly easy to check that eight do not tile any rectangle. The surprising exception is the Y-pentomino. This pentomino, which can be seen in Figure 6.2 tiling an infinite strip, does tile a rectangle. A hard part of the problem is to find a rectangle that can be tiled with the Y-pentomino in the first place. This part is given away in the statement of our first problem below. Therefore, for a very challenging puzzle, stop reading now and find a square tiled with the Y-pentomino. The Y-pentomino is unusual in that the smallest number of copies that tile a rectangle is not 1, 2, or 4, as is most often the case for the polyominoes that do tile a rectangle. Tiling rectangles with the Y-pentomino provides many demanding puzzles. For example, you might want to try some of the following rectangles, each of which has a tiling: 10×14, 10×16, 9×20, 14×15, 11×20, 15×16, 9×30, 15×22,

FIGURE 6.1

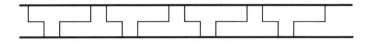

FIGURE 6.2

and 22×25. Copies of a rectangle pasted end-to-end tile an infinite strip and copies of this infinite strip can be pasted together to make infinitely many different tilings of the plane. The polyominoes that tile a rectangle tile the plane in infinitely many ways.

Odd Pentominoes

The pentominoes are remarkable in that three of the twelve are *odd*. That means that there is an odd integer such that this many copies can be put together to form a rectangle; by convention, rectangles are not odd. From Figure 3.9 we see that the L-tromino is odd. It follows from the discussion above that the odd pentominoes must be the L-pentomino, the P-pentomino, and the Y-pentomino. Twenty-seven copies of the L-pentomino tile a 9×15 rectangle. Twenty-one copies of the P-pentomino tile a 7×15 rectangle. Finally, forty-five copies of the Y-pentomino tile a 15×15 square. If you want a puzzle that is guaranteed to induce insanity, then finding a tiling of the 15×15 square with the Y-pentomino is right up your alley. A glutton for punishment may wish to go on to do battle with the 25×27 rectangle or the 27×35 rectangle.

Problem 6.1

Tile a 5×10 rectangle with the Y-pentomino.

Problem 6.2

Tile a 7×15 rectangle with twenty-one copies of the P-pentomino.

Problem 6.3

Find a fault-free tiling of a 15×15 square with the P-pentomino.

Problem 6.4

Tile a 9×15 rectangle with twenty-seven copies of the L-pentomino. Extend this tiling to a tiling of a 15×15 square.

Problem 6.5

Is there a fault-free tiling of a square with the L-pentomino?

Problem 6.6

Tile a 15×15 square with forty-five copies of the Y-pentomino.

Since four of the pentominoes tile a rectangle, those four are also reptiles. Of course, the I-pentomino is rep 4. Since each of the L-pentomino and the P-pentomino tiles a 2×5 rectangle and since the Y-pentomino tiles a 5×10 rectangle, each of these three pentominoes tiles a 10×10 square and is therefore rep 100. Only one of Problems 6.7, 6.8, and 6.9 has an easy solution. It is easy to check that none of the other eight pentominoes is a reptile. Every known polyomino reptile tiles some square. Hence, the answer to Problem 6.10 is unknown.

Problem 6.7

What is the smallest integer k such that the L-pentomino is rep k?

Problem 6.8

What is the smallest integer k such that the P-pentomino is rep k?

Problem 6.9

What is the smallest integer k such that the Y-pentomino is rep k?

Problem 6.10

Is there is a polyomino reptile that does not tile a rectangle?

Four of the eight pentominoes that do not tile a rectangle do tile a strip having parallel lines as edges. You can form infinitely many tilings of the plane with each of these pentominoes by pasting together copies of one of the strips. Of the next four problems, the one involving the V-pentomino is the most challenging.

Problem 6.11

Show that the F-pentomino tiles an infinite strip having parallel lines as edges.

Problem 6.12

Show that the N-pentomino tiles an infinite strip having parallel lines as edges.

Problem 6.13

Show that the V-pentomino tiles an infinite strip having parallel lines as edges.

Problem 6.14

Show that the W-pentomino tiles an infinite strip having parallel lines as edges.

Copies of the two infinite strips in Figure 6.3 can be pasted together to form infinitely many tilings of the plane with the T-pentomino. We

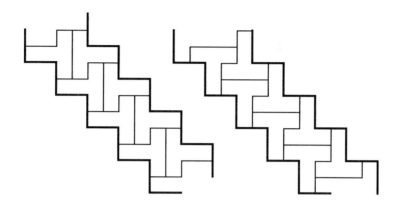

FIGURE 6.3

are not claiming that the two strips are congruent, only that they stack with each other.

Copies of the two infinite strips in Figure 6.4 can be pasted together to form infinitely many tilings of the plane with the U-pentomino.

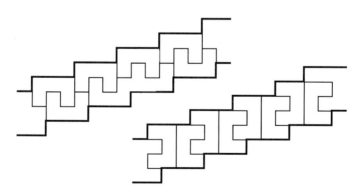

FIGURE 6.4

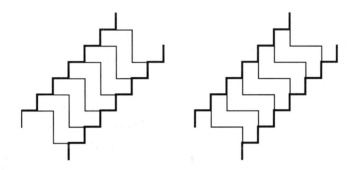

FIGURE **6.5**

Copies of the two infinite strips in Figure 6.5 can be pasted together to form infinitely many tilings of the plane with the Z-pentomino. Note that in Figure 6.5 we used both "sides" of the Z-pentomino.

Except for the X-pentomino, which we will discuss after the statement of Problem 6.15, and the Z-pentomino, infinitely many tilings of the plane are possible "without turning over the polyomino" for each n-omino with $n \leq 5$. It will be a review for you to check that this is true; the solutions that will be given later to the preceding four problems and the infinite strip tiled by the Y-pentomino in Figure 6.2 should cover any doubtful case. The Z-pentomino is remarkable in that there are exactly six tilings of the plane that can be constructed without turning over the pentomino. Tilings are considered to be the same if they are congruent, that is, if by rotating, sliding, and flipping you can fit one tiling on top of the other to have the tilings coincide. If one figure is obtained from another by sliding and rotating only, without any flipping, then the figures are said to be *directly congruent* to each other. In Figure 6.6, copy B of a Z-pentomino is directly congruent to copy A, but neither copy

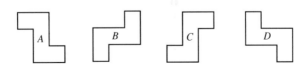

FIGURE **6.6**

C nor copy D is directly congruent to copy A. We say copies A and B show the "front side" of the Z-pentomino in the figure, while copies C and D show the "back side."

Problem 6.15

Find six tilings of the plane with directly congruent copies of the Z-pentomino.

The X-pentomino is remarkable in that it is the first of the polyominoes that is *monomorphic*. This means there is exactly one way to tile the plane with the polyomino. The most familiar polygon that shares this property is not a polyomino but the regular hexagon. Once one copy of the regular hexagon is placed in the plane, the placing of all other copies of the hexagon is determined. This familiar edge-to-edge tiling is shown in Figure 6.7. Tiling with the X-pentomino is only slightly different. Placing copy A down and selecting the slot that is the concave corner of copy A indicated by the square dot in Figure 6.8, we have two choices in the way we can fill this slot with a second copy of the X-pentomino. Once one of the two configurations in Figure 6.8 is chosen,

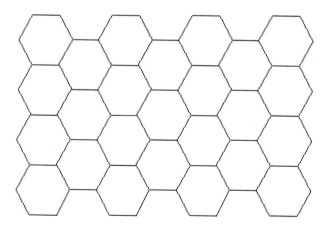

FIGURE 6.7

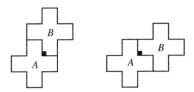

FIGURE **6.8**

there are no more choices in completing that configuration to a tiling of the plane with the X-pentomino. The tilings of the plane are indicated in Figure 6.9. Since the two configurations in Figure 6.8 are congruent, we should not be surprised to find that the tilings determined by these configurations are also congruent. Each of the tilings indicated in Figure 6.9 is obtained from the other by a reflection. Since these tilings of the plane are congruent, they are not considered to be different. This seems reasonable, since if you were paying a designer for designs you would probably be upset to be billed separately for the designs in Figure 6.9.

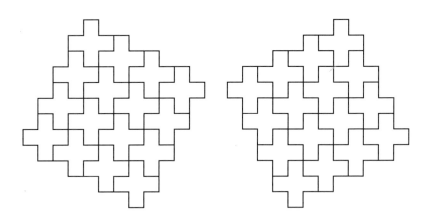

FIGURE **6.9**

Backtracking

We have already seen that the Z-pentomino tiles the plane in infinitely many ways, if as usual both sides of the pentomino are allowed in the tilings. We will investigate the problem of tiling the plane with only directly congruent copies of a Z-pentomino. It might be mentioned that the restriction of not being allowed to turn over a tile is realistic, since the tiles used in the everyday world have a front side that is finished and that is distinct from the rough back side. Even so, the motivation here is more to illustrate the technique of backtracking , a method that assures all possible cases in a complicated analysis have been considered. You can skip to the next section now and return to backtracking when or if you feel the need. However, it will be worth the effort of following the lengthy example of finding all tilings of the plane with directly congruent copies of the Z-pentomino in order to become familiar with this very powerful mathematical method. It is not enough to do Problem 6.15. Finding six tilings of the plane with directly congruent copies of the Z-pentomino does not prove there is not a seventh lurking out of sight. The if-I-can't-find-any-more-then-there-aren't-any-more argument is persuasive only if put forward by an omniscient being.

FIGURE **6.10**

To best follow the lengthy investigation, it is desirable to have about thirty copies of the Z-pentomino with the six projecting corners on the front side labeled 1, 2, 3, as in Figure 6.10. These copies can be cut from paper and labeled. All copies are assumed to be labeled in this fashion, even though they are not labeled in the figures that follow. The redundancy in labeling is due to the $180°$ rotation symmetry of the Z-

pentomino. Although we have distinguished between the front and the back, we do not distinguish the corners with the same numeral.

We begin with copy A of a Z-pentomino and designate a slot to be filled with corner 1 of copy B. The slot in copy A is indicated in Figure 6.11 by the square dot. After finding all the possible tilings that contain the initial configuration AB of Figure 6.11, we must backtrack to consider in turn each of the other two possibilities for filling the slot in Figure 6.11. Whenever a dead end or a tiling is reached in the analysis, we must backtrack to the last stage where a decision among two or more possibilities was made and consider the next possibility. This backtracking must be continued until all possibilities have been exhausted. Our situation is very much like traversing a labyrinth, where at each intersection there are three possible paths. One strategy is to label—say from left to right—the alternative paths 1, 2, and 3 from each intersection as the intersection is approached for the first time. We always take the paths from an intersection in numerical order until a dead end is reached, in which case we backtrack to the last intersection where a choice was made and take the next alternative. For a labyrinth or maze, this strategy dictates that you touch the left wall with your left hand and continue walking without your left hand losing contact with the wall. Repetition of this backtracking assures your escape from the labyrinth or else your return to the starting point after having traversed all the paths.

In the traversal of our labyrinth, we are only at the beginning, having configuration AB as our initial stage. We next pick a slot in configuration AB to be filled with the third copy. The only two possible ways of placing a third copy in this slot are shown in Figure 6.12; the apparent third possibility is discarded immediately as the indicated slot cannot be filled with corner 3 of a copy without overlapping. From the choice of letters, you may have guessed we are going to consider configuration ABC, which has corner 2 in the slot, before configuration ABD, which has corner 1 in the slot. If you are doing this analysis for the first time you do not realize the short cuts that can arise by deviating from the natural order 1, 2, 3. Unlike the labyrinth situation, we learn some tricks as we go along. Choosing corner 2 before corner 1 does not in any way relieve us from the necessity of coming back to consider the alternative. It is only in retrospect that we will be able to see that this is a clever choice. Trust me.

We have the configuration ABC, as in Figure 6.12 and as repeated in Figure 6.13. Choosing the indicated slot for copy D in Figure 6.13, we see that there is only one possible corner for D that fills the slot without overlapping. This is nice because we will never have to backtrack to this D again. In other words, the position of D in Figure 6.13 is determined by the configuration ABC. With planning and good luck, this situation happens a lot. Picking the slots to be filled next at a given stage is something of an art, which improves with practice. Now, copies E, F, G, and H are likewise determined by the configuration. These and the choice of corner 1 to fill the indicated slot with copy I are shown in Figure 6.13. There are two choices for the corners of copy I. The choice of corner 1 is shaded in Figure 6.13. We will have to backtrack to this configuration again and consider corner 2. For now we have corner 1 and then copies J, K, and L are determined in the figure. However, we now have a slot

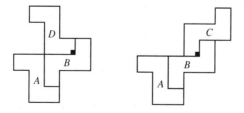

FIGURE 6.12

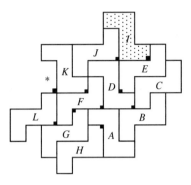

FIGURE 6.13

that cannot be filled with any corner of a copy. We have hit a dead end. We must backtrack to our last choice. This was the placing of copy I.

The only remaining possibility for copy I is indicated in Figure 6.14. At this point we know that the configuration ABC determines configuration $ABCDEFGHI$. Further, since configuration DEI is congruent to configuration ABC, as configuration ABC determines D, E, F, G, and I, so configuration DEI determines J, K, L, M, and N. We are on a roll. Configuration JKN is also congruent to configuration ABC. The motion that takes configuration ABC to configuration DEI takes configuration DEI to configuration JKN and so on. The tiling is forced to repeat in this manner. Since configuration DFG is also congruent to configuration ABC, the pattern is forced to repeat in the opposite direction in the same manner. The tiling of an infinite strip, part of which is highlighted in the lower left part of Figure 6.14, is now determined. Since the strip has 180° rotation symmetry, then the tiling of the entire plane is determined if the tiling on all of one side of the strip is determined. Copies O and P in the figure are determined next. Since configuration PIE is congruent to configuration ABC, then Q, R, S, and T are determined, as configuration ABC respectively determines copies M, F, G, and H. Now, configuration ORS is also congruent to configuration ABC and has the same aspect. The tiling is forced to repeat in the direction that takes configuration ABC to configuration ORS. Note that the highlighted infinite strip in the figure must also repeat but that

FIGURE 6.14

successive repetitions partially overlap each other. A repeating pattern that does not overlap itself is shown in the lower right part of the figure. The tiling on one side of the indicated infinite strip is determined, and thus the tiling of the entire plane is determined. Our first tiling of the plane with directly congruent copies of the Z-pentomino is indicated in Figure 6.14. Exactly one tiling is determined by configuration ABC.

While traversing a labyrinth, there is no possibility of learning a rule such as "never take three successive left-hand turns," just because several of these decisions have lead to dead ends. However, in our situation, we can learn new rules as we go along. For example, configuration ABC is forbidden in our investigation from now on. We can do this because if configuration ABC pops up we can say either one of two things must happen. Either we will necessarily end up with our first tiling again or else we will run into a dead end because of some incompatibility. We are now not interested in either case.

We backtrack to the last place a choice was made. This was at C. We retire the letter C and return to the configuration ABD of Figure 6.12 and repeated in Figure 6.15. Since we cannot have three copies congruent to the configuration ABC, then copy E in Figure 6.15 is determined. Therefore, at this point we know that configuration AB forces configuration $ABDE$. The slot indicated in the figure can be covered with two possible corners, one of which is corner 1 of copy F shown in Figure 6.16. Copy G is determined. As configuration AB forces configuration $ABDE$, so configuration FG forces configuration $FGHI$. Copy J is now determined, as is copy K. As configuration AB forces configuration $ABDE$, so configuration JK forces configuration $JKLM$. Next, copies N, O, P, Q, and R are determined in turn. Since each of configuration $GHIFN$ and configuration $IFGHB$ is congruent to configuration $ABDEF$, then we must have an infinite strip tiled with Greek crosses that are themselves tiled with four Z-pentominoes. (A Greek cross has the same shape as the X-pentomino.) The bottom two of the four nonoverlapping Greek crosses that are highlighted in Figure 6.16 are part of this infinite strip. Since this strip has $180°$ rotation symmetry, the tiling of the entire plane is determined if a tiling of all of one side of the strip is determined. Now, since configuration $ROPQM$ is congruent to configuration $ABDEF$, we see that copies of the infinite

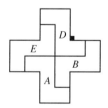

FIGURE 6.15

FIGURE 6.16

strip must be stacked together to cover the plane. Our second tiling of the plane with directly congruent copies of the Z-pentomino is determined as indicated in Figure 6.16.

There are several ways to look at our second tiling. One way is as the tiling from the right side of Figure 6.9 with the tiles subdivided. Another way is to view the tiling as two copies of the tiling from the right side of Figure 6.9, with one tiling superimposed on the other. A third way is as a repetition of the tiled ratchet figure highlighted in the lower right corner of Figure 6.16. Considering Figures 6.9 and 6.16 together, we might anticipate the existence of another tiling, which will not turn up until Figure 6.21. We do not list that tiling now because that would distract us from the path dictated by the backtracking method. We are, however, confident that our method will turn up the anticipated tiling before we are done. We are not looking for some tilings with directly congruent copies of the Z-pentomino but following a path that assures us of all such tilings.

We backtrack to the last place a choice was made. This was at F. We retire the letter F and go to the configuration $ABDEG$ of Figure 6.17. Since G is now determined in the figure, so are H, I and J by the symmetry of a 90° rotation. We now know that configuration AB determines configuration $ABDEGHIJ$. It would be false reasoning to jump to the conclusion that any new tiling containing configuration AB must have a 90° rotation symmetry.

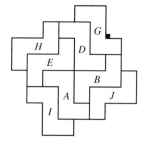

FIGURE 6.17

There are three ways to cover the slot shown in Figure 6.17. One of these ways is with copy K as shown in Figure 6.18. Again we are knowingly deviating from taking the corners in numerical order. Copies L, M, and N in the figure are determined in turn. As configuration AB determines the relative placing of copies D, E, G, and H, so configuration KN determines copies O, P, Q, and R in the figure. Next, copies S and T are determined in turn. Since configuration TS is congruent to configuration AB, then copies U, V, W, X, and Y are determined. Now copies Z and Z' are determined in turn, and, with these, copies K' and K'' are determined. Each of configuration $VUTSYK'$, configuration

FIGURE 6.18

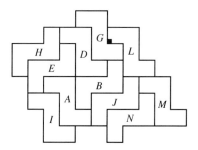

FIGURE 6.19

$KNOPQK''$, and configuration $OPKNGD$ is congruent to configuration $ABDEGK$. From this it follows that a parade of "framed" Greek crosses marching in a two-way infinite progression is forced to repeat itself over and over again to cover the plane. The remainder of the tiles in the tiling of the plane indicated in Figure 6.18 are determined. Note that this tiling has a 180° rotation symmetry but does not have a 90° rotation symmetry. Three down and three to go.

Maybe it is time for a break. One thing about backtracking arguments is that they can be left at a point where one is returning to the last place a choice was made and later picked up at that point without retracing all that went before. We are at such a point now.

We backtrack to the last place a choice was made. This was at K. We retire the letter K and go to the configuration $ABDEGHIJL$ of Figure 6.19. A short argument, much like that dealing with Figure 6.13, shows that copy M is determined in Figure 6.19. Copy N is then determined. Since configuration $EABDJN$ of Figure 6.19 is congruent to configuration $ABDEGK$ of Figure 6.18, we see that Figure 6.19 determines the same tiling that we just considered. From our perspective, we have the preceding tiling rotated 90°. We are at another dead end. Still three to go, but it's all downhill from here.

We backtrack to the last place a choice was made. This was at L. We retire the letter L and go to the configuration $ABDEGHIJM$ of Figure 6.20. Since copy M is now determined by the configuration AB, then copies N, O, and P are also determined by symmetry. At this point we have that configuration AB determines all of configu-

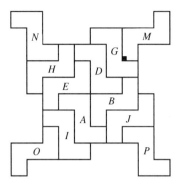

FIGURE 6.20

ration $ABDEGHIJMNOP$, that is, all of Figure 6.20. Further, since this larger configuration contains so many configurations that are themselves congruent to configuration AB, we see that configuration AB now determines the rest of the tiling of the plane that is indicated by Figure 6.21. We have our fourth tiling of the plane with directly congruent copies of the Z-pentomino.

We backtrack to the last place a choice was made. We have to go back all the way to copy B. All the tilings of the plane with direct copies of the Z-pentomino that contain the configuration AB of Figure 6.22 have been detected. Henceforth, the configuration AB of Figure 6.22 is forbidden. In our backtracking, we have yet to consider the configurations AN and AO of Figure 6.23. We note the combination of these configurations that is shown in Figure 6.24 is impossible, since the slot indicated in the figure forces the now forbidden configuration AB. It quickly follows that there are only two possible tilings, one arises from configuration AN and the other from configuration AO. These tilings are indicated in Figures 6.25 and 6.26. These are the fifth and sixth tilings of the plane with direct copies of the Z-pentomino. Further, we are finally at the end of the backtracking, since we have considered all possibilities of filling the indicated slot in the initial copy A.

There are six tilings of the plane with directly congruent copies of the Z-pentomino and no more. The figures alone show that there are

FIGURE 6.21

at least six tilings. The backtracking argument not only provided the six tilings but also proved that there are no more than these six tilings.

Backtracking arguments can be much more complicated than the argument just completed. Here, we were lucky enough never to have more than four choices piled up at one time. Although lengthy and tedious, the technique is still almost as simple-minded as traversing a labyrinth by walking without taking your left hand from the wall.

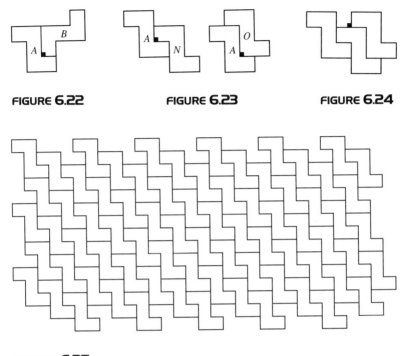

FIGURE **6.22**   FIGURE **6.23**   FIGURE **6.24**

FIGURE **6.25**

FIGURE **6.26**

Polymorphic and Polypoic Polygons

With the introduction of new ideas comes the introduction of new words to describe these ideas. A polyomino is said to be m-*morphic* if congruent copies of the polyomino tile the plane in exactly m noncongruent ways. A polyomino is said to be p-*poic* if directly congruent copies of the polyomino tile the plane in exactly p noncongruent ways. For positive integers m and p, an m-morphic polyomino is *polymorphic,* and a p-poic polyomino is *polypoic.* (The accented syllable of *polypoic* is the first syllable of *poet.*) The X-pentomino is 1-morphic. We have just finished the long backtracking argument that shows that the Z-pentomino is 6-poic. As with the names of the polyominoes, the Greek prefixes are used for small values of m and p as in the following list.

1-morphic or monomorphic,	1-poic or monopoic,
2-morphic or dimorphic,	2-poic or dipoic,
3-morphic or trimorphic,	3-poic or tripoic,
4-morphic or tetramorphic,	4-poic or tetrapoic,
5-morphic or pentamorphic,	5-poic or pentapoic,
6-morphic or hexamorphic,	6-poic or hexapoic,
7-morphic or heptamorphic,	7-poic or heptapoic,
8-morphic or octamorphic,	8-poic or octapoic,
9-morphic or enneamorphic,	9-poic or enneapoic,
10-morphic or decamorphic,	10-poic or decapoic.

After the first four, these are the same prefixes that we are familiar with from the names of the polygons: pentagon, hexagon, heptagon, octagon, enneagon, and decagon. Two of our results can be stated as follows. The X-pentomino is monomorphic. The Z-pentomino is hexapoic.

Definitions are abbreviations. It is simply easier to say, "the Z-pentomino is hexapoic" than to say "directly congruent copies of the Z-pentomino tile the plane in exactly six noncongruent ways." What at first seems unnecessary jargon will become welcome terminology with use, and we will use the terminology introduced above frequently.

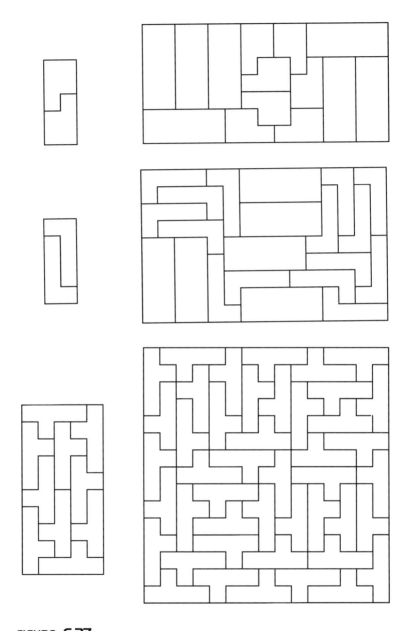

FIGURE 6.27

For those who take pleasure in words for their own sake, we introduce four more technical terms. These need not be mastered as they will always be explained each time they appear. A 0-morphic polyomino does not tile the plane at all and is said to be *gymnomorphic*. Similarly, a 0-poic polyomino is *gymnopoic*. The Greek prefix here means naked, as in the word gymnasium. (In ancient Greece, clothes were not worn during athletic exercise.) At the other extreme, since hyper means excessive in Greek, we may call a polyomino that tiles the plane in infinitely many noncongruent ways *hypermorphic*. Similarly, if directly congruent copies of a polyomino tile the plane in infinitely many noncongruent ways, we say the polyomino is *hyperpoic*.

Figure 6.27 shows that the three pentominoes that tile a rectangle and that are not rectangles themselves are odd. The 7×15 rectangle tiled with the P-pentomino in the figure can be extended to a 15×15 square by the addition of twelve 2×5 rectangles all having the same aspect as the 2×5 rectangle in the lower left corner of the 7×15 rectangle in Figure 6.27. This corner rectangle and the one adjacent in this 15×15 square form one of the 4×5 rectangles that can be replaced by the tiled 4×5 rectangle of Figure 6.28 to form a fault-free tiling of a 15×15 square with the P-pentomino.

FIGURE 6.28

Figure 6.29 shows that the P-pentomino is rep 4; the P-pentomino is rep 4, as is shown by Figure 6.29.

Figure 6.30 shows that directly congruent copies of each of the F-pentomino, the N-pentomino, the V-pentomino, and the W-pentomino tile an infinite strip with parallel lines as edges.

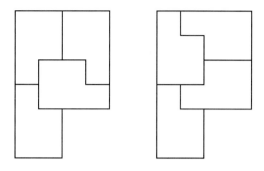

FIGURE 6.29

FIGURE 6.30

THE HEXOMINOES

Well there they are, all thirty-five of them in their best black and white clothes. Let's see, $35 \times 6 = 210$. So, if the hexominoes are to form a rectangle, then the sides must be factors of 210. Since the 2×105 rectangle is out of the question, the candidates for the rectangles are the 3×70, the 5×42, the 6×35, the 7×30, the 10×21, and the 14×15 rectangles.

However, you may recall that Problem 1.5 asks you to show that the hexominoes do not form any rectangle. We will solve that problem now. The hexominoes come in two flavors, which we will call "balanced" and "unbalanced" for the moment. The first twenty-four shown above are called balanced because no matter where they are placed on a checkerboard each one covers three black squares and three white squares. The remaining eleven are unbalanced, and each of them will cover either two black and four white squares or else four black and two white squares wherever placed on a checkerboard. Each unbalanced hexomino covers an even number of black squares. It is not important that the number of unbalanced hexominoes is odd. It is important that the number of balanced hexominoes is even though. No matter how the unbalanced hexominoes are distributed among those that cover two or four black squares, the total number of black squares that are covered by all of the hexominoes on an enlarged checkerboard is even. Since 210 is even, the number of black squares to be covered by all of the hexominoes formed into a rectangle with a checkerboard coloring is the same as the number of white squares to be covered and this number is 105, which is certainly an odd integer. Since 105 is not even, we will have to agree that all the hexominoes cannot be formed into one rectangle.

We can apply the checkerboard coloring to any figure that might be presented to be covered by all of the hexominoes. A slightly deeper analysis shows that the the difference between the number of black and white squares in the presented figure of 210 squares must be 2, 6, 10, 14, 18, or 22 or else, no dice. It may or may not be possible to cover a figure satisfying this criterion with all of the hexominoes. What is certain is that the criterion must be met if there is to be any possibility. The analysis goes as follows. Suppose there are x hexominoes that cover two black squares and four white squares in the presented figure. Then there are $11 - x$ that cover four black squares and two white squares. Of course, there are the 24 balanced hexominoes that cover three squares of each color. Hence we have $2x + 4(11 - x) + (24)(3)$ black squares and $2(11 - x) + 4x + (24)(3)$ white squares covered. The difference is $\pm(4x - 22)$ where x is an integer from 0 to 11. The special case of the rectangle above follows from the fact that 0 is not among the possible differences 2, 6, 10, 14, 18, and 22.

Hexomino Puzzles

The grand staircase in Figure 7.1 has one hexomino placed "to echo the outline," as *The Fairy Chess Review* put it. The problem is to use the rest of the hexominoes to finish covering the figure. Each of the fourteen rows is fifteen squares long. Therefore, a checkerboard coloring of the figure will show a difference of 14 between the number of black and white squares.

Problem 7.1

Cover the grand staircase of Figure 7.1 with all of the hexominoes, where one hexomino has already been centrally placed to echo the outline.

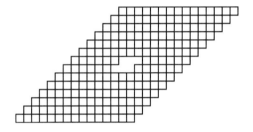

FIGURE 7.1

Problem 7.2

With all of the hexominoes, cover the 15 × 17 rectangle with a Greek cross having bars of width 3 removed from the center, as shown in Figure 7.2.

Problem 7.3

With all the hexominoes, cover the 5 × 45 rectangle with fifteen unit squares removed as shown in Figure 7.3.

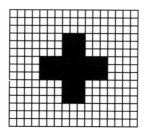

FIGURE 7.2

FIGURE 7.3

The next two of our problems from *The Fairy Chess Review* reflect the interest of the journal. You might copy the outlines for later use as a puzzle. Having seen a solution once does not decrease the enjoyment of reconstructing these puzzles at a later date.

Problem 7.4

Cover the black rook figure in Figure 7.4 with the twelve pentominoes. Then cover the remaining part of the 15×18 rectangle with the thirty-five hexominoes.

Problem 7.5

Cover the rook figure of Figure 7.5 with the hexominoes.

Problem 7.6

Form the twenty-four balanced hexominoes into a 12×12 square having a 6×6 square in one corner.

As well as the six problems above and others, *The Fairy Chess Review* gave the following three that are mentioned here in passing. Cover

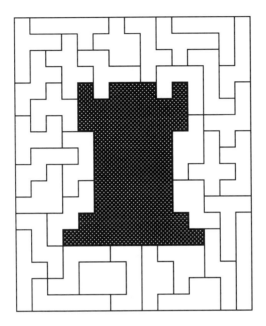

FIGURE 7.4

a 5 × 84 rectangle with a double set of hexominoes. Cover a 15 × 28 rectangle with a double set of hexominoes. With a double set of the hexominoes, cover a 26 × 26 square having a centrally placed 16 × 16 square removed.

Since $35 = 1^2 + 3^2 + 5^2$, it makes arithmetical sense to try to divide the hexominoes into sets of one hexomino, nine hexominoes, and twenty-five hexominoes such that each of the sets of nine and twenty-five can be arranged to form a figure that is similar to the single hexomino. As the next problem indicates, there is at least one solution. Are there any more?

Problem 7.7

With the cross of Figure 7.6 removed from the set of hexominoes, divide the remaining hexominoes into sets of nine and

FIGURE 7.5

twenty-five hexominoes such that each of these two sets can be arranged to form a figure similar to the cross.

FIGURE 7.6

Two observations about the cross of Figure 7.6 will suggest two more problems. First, you can cut a cube along seven of its edges in such a way that the flattened cube surface presents the cross of the figure. Second, it is possible to trace this cross without lifting the pencil from the paper and without traversing an edge of any unit square more than once.

Problem 7.8

Which of the hexominoes can be formed in one piece by cutting selected edges of a cube?

Problem 7.9

Which of the hexominoes, shown with their individual unit squares, can be traced without lifting the pencil from the paper and without traversing any segment more than once?

Tiling with Hexominoes

FIGURE 7.7

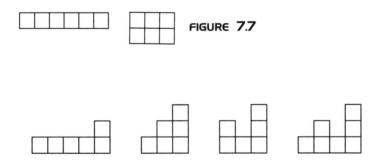

FIGURE 7.8

Two of the hexominoes are rectangles. See Figure 7.7. Two copies of any one of the four hexominoes in Figure 7.8 easily form a rectangle. Two copies of the hexomino in Figure 7.9 easily form a rectangle also. However, this hexomino has the additional property of being odd. Recall that a polygon is odd if for some odd integer, greater than 1, that many copies form a rectangle. In this case, eleven copies form a rectangle. Is there another nonrectangular polyomino such that n copies form a rectangle where n is odd and $n \leq 11$? The answer to this question (Problem 7.11) is unknown. It takes four copies of the hexomino in Figure 7.10 to form a rectangle. A polyomino that tiles a rectangle usually does so with one, two, or four copies. The hexomino in Figure 7.11 is unusual in that it does tile a rectangle but it takes eighteen copies to do so.

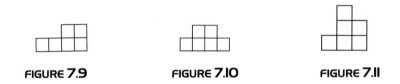

FIGURE 7.9   FIGURE 7.10   FIGURE 7.11

Problem 7.10

Form a 6×11 rectangle with eleven copies of the hexomino in Figure 7.9.

Problem 7.11

Is there another nonrectangular polyomino such that n copies form a rectangle where n is odd and $n \leq 11$?

Problem 7.12

Form a 9×12 rectangle with eighteen copies of the hexomino in Figure 7.11.

The figures above show all the hexominoes that tile a rectangle with one exception. After nearly twenty years of effort by scores of amateur and professional mathematicians, the hexomino in Figure 7.12 was finally found to tile a rectangle. This result was obtained in 1987 on a personal computer by Karl Dahlke, who is blind. You will probably be the first on your block to do Problem 7.13, but you might be the first on your planet to find out whether this hexomino is odd.

Problem 7.13

Form a rectangle with ninety-two copies of the hexomino in Figure 7.12.

Problem 7.14

Is the hexomino in Figure 7.12 odd?

FIGURE 7.12

Recall that a polyomino is p-poic if exactly p noncongruent tilings of the plane can be made with directly congruent copies of the polyomino. Three of the hexominoes are polypoic.

Problem 7.15

Show that the hexomino in Figure 7.13 is dipoic.

FIGURE 7.13   FIGURE 7.14

Problem 7.16

Show that the hexomino in Figure 7.14 is tripoic.

Problem 7.17

Show that the hexomino in Figure 7.15 is tripoic.

All thirty-five of the hexominoes are hypermorphic (i.e., tile the plane in infinitely many noncongruent ways). Thirty-two of the hexominoes are hyperpoic (i.e., tile the plane in infinitely many noncongruent ways with directly congruent copies). It is a good exercise to run through the hexominoes to check these two statements. Further, it is not as hard as it might seem at first, since twenty-five tile infinite strips with parallel lines as edges. Of these twenty-five hexominoes, you need to use both sides of the hexomino only for the most challenging one, which is shown in Figure 7.16. For the remaining cases, you can find infinite strips tiled with the given hexomino or with directly congruent copies of the hexomino, as the case may be, such that the strips can be stacked to form infinitely many different tilings of the plane. Backtracking is not needed for this but is needed for Problems 7.15, 7.16, and 7.17. Remember that finding the indicated number of tilings is not enough to solve any of these problems; backtracking is needed to prove that there are no more tilings than the indicated number. A backtracking exercise that is a little more involved than that required for these three problems is stated as our last numbered problem for this chapter.

FIGURE 7.15   FIGURE 7.16

Problem 7.18

Show that the hexomino in Figure 7.16 does not tile a rectangle.

Puzzle Solutions

A solution to the grand staircase of Problem 7.1 is shown in Figure 7.17. A solution to the Greek cross of Problem 7.2 is shown in Figure 7.18. A

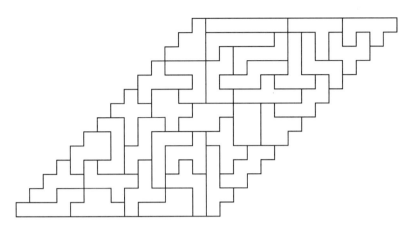

FIGURE 7.17

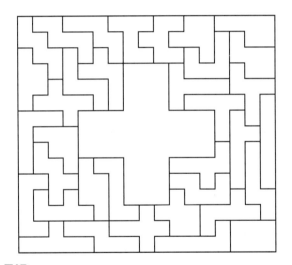

FIGURE 7.18

solution to the perforated rectangle of Problem 7.3 is shown in Figure 7.19.

In Figure 7.20, the balanced hexominoes are arranged into a 12×12 square having a 6×6 square in one corner. This solves Problem 7.6.

Problem 7.8 is terribly hard as stated. Because the cube has so many symmetries, there are several ways to cut the edges to produce the same

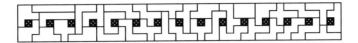

FIGURE 7.19

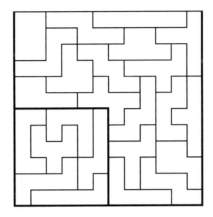

FIGURE 7.20

hexomino. Rather than starting with a cube and trying to get to a hex-omino by cutting selected edges, reverse the process. Start with a hex-omino and try to fold it into a cube. The same problem when restated is very much easier to solve. Which of the hexominoes can be folded along the edges of their unit squares to form a cube? Examine the thirty-five hexominoes one by one to find which fold into a cube. There are eleven. Do these eleven hexominoes form a rectangle?

Graphs

The argument concerning which hexominoes can be traced without lift-ing the pencil and without traversing a segment more than once applies to any figure that consists of a finite set of vertices and a finite number of curves drawn to join pairs of these vertices. The curves joining the vertices are customarily called "edges" in this context. Edges joining a vertex with itself are allowed and are no problem since they can easily be traced under the given conditions. Also, two vertices can be joined by more than one edge without causing any problem. (Such sets of vertices and edges are called *graphs*. Strange, but true.) If we are to have any hope of traversing the figure, then the figure must be connected in the sense that you can trace some curve, consisting of a sequence of edges, from any vertex to any other vertex. The polyominoes with their individual squares satisfy this requirement.

A vertex in one of these figures is "odd" or "even" as it has an odd or even number of edges connected to it. Our first observation is that the number of odd vertices is necessarily even, since the sum over all the vertices of the number of edges connected to that vertex equals twice the total number of edges and this number is necessarily even. Suppose the figure can be traversed under the stated conditions. Except possibly for the first and last vertices visited, there must be a matching pair of edges through any vertex for each time the tracing passes through that vertex, one edge traced coming into the vertex and one edge traced leaving the vertex. Thus, since all the edges are traced, then, except possibly for both the first and last vertices, each vertex must be even. In other words, a figure that can be traced under the stated conditions can have at most two odd vertices. If there are two odd vertices, then the tracing must start at one odd vertex and end at the other odd vertex. If there are no odd vertices, then you will end up wherever you started.

We have found a necessary condition for our problem. A figure of vertices and edges to be traced can have at most two odd vertices. We want to argue that this condition is also sufficient. That is, if a figure has at most two odd vertices then the figure can be traced under the stated conditions. There are two cases, depending on the presence of odd vertices. We start with the case where all vertices are even. In tracing the figure, it makes no difference where we start. Having begun, as we approach any vertex in our tracing, we know there must be an escape edge to trace and head for another vertex until we have returned to the starting vertex and have no escape. With good luck, we have made a grand tour and are done. Without the good luck, our grand tour left some edges untraced. In this case, we need a side trip that begins at a vertex already visited and connected to one of these untraced edges. Since this vertex is even, an edge traced leaving the starting point of the side trip leaves an untraced edge available to be traced upon returning.

Since all the vertices visited in between on the side trip are also even, matching the edge traversed to arrive at each of these vertices always leaves an edge free to be traversed and provide escape from that vertex. Having thus necessarily arrived back at the starting point of the side trip, we then incorporate the side trip into our grand tour. If bad luck is still dogging us and there are still edges untraced, we play the side trip game again and again until all the edges are covered. We know the process must eventually terminate with all edges traversed exactly once since the figure is connected with a finite number of edges. For the case where there are two odd vertices, we add an edge between these two vertices to reduce the problem to the former case where all vertices are even. In the tracing, we simply agree to trace the added edge first. Then, ignoring this path along the added edge, we have a path that starts at one odd vertex and returns to the other, as expected.

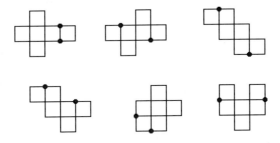

FIGURE 7.21

For our Problem 7.9, the hexominoes that can be traced under the stated conditions are shown in Figure 7.21. Each of the six hexominoes has two odd vertices, which are indicated in the figure. A tracing of the hexomino must start at one of these vertices and end at the other. The X-pentomino has all even vertices. You can easily create such figures, with or without two odd vertices, by making a drawing without lifting your pencil from the paper.

Three Remarkable Rectangles

The tiling of a rectangle with eleven copies of the odd hexomino is shown in Figure 7.22. This solves Problem 7.10.

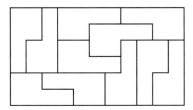

The smallest rectangle tiled with the hexomino shown in Figure 7.23 is given in that figure. This solves Problem 7.12.

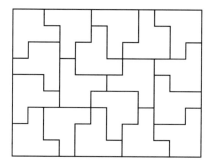

The smallest rectangle tiled with the hexomino shown in Figure 7.24 is the 23 × 24 rectangle given in that figure. This solves Problem 7.13.

FIGURE 7.24

Polypoic Hexominoes

In Problem 7.15 we are asked to show that the hexomino in Figure 7.25 is 2-poic. We will only outline the backtracking that proves this. With hindsight, it is seen that the labeling of the corners and the selected slot, shown in Figure 7.25, are convenient. Due to the 180° rotation symmetry of the hexomino, we need label only four corners. Placing the corner labeled 1 of copy B into the indicated slot of copy A immediately leads to an impossibility. The corner labeled 2 will not fit into the slot of copy A. Placing the corner labeled 3 into the slot of copy A then leads to

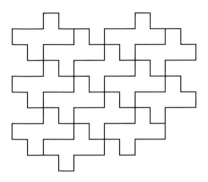

FIGURE **7.27**

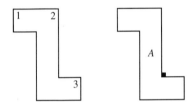

FIGURE **7.28**

A now leads to a tiling that is suggested by Figure 6.25. The hexomino is tripoic.

The hexomino in Figure 7.28 tiles an infinite staircase in two ways, and so the hexomino tiles the plane in infinitely many ways, formed by stacking the staircases. We do not need infinite strips to confirm that the hexomino in Figure 7.30 also tiles the plane in infinitely many ways. The hexomino tiles an *I*-shaped 12-omino that is easily seen to tile the plane. This tiling of the plane with the 12-omino can be changed to a tiling with the hexomino by dividing each 12-omino into two copies of the hexomino. Since the hexomino tiles the 12-omino in two ways that are not directly congruent, it follows that there are infinitely many choices to make in dividing the 12-ominoes into two copies of the hexomino. This provides infinitely many tilings of the plane with the hexomino.

In Problem 7.17 we are asked to show that the hexomino in Figure 7.31 is 3-poic. We will only outline the backtracking that proves this. With hindsight, it is seen that the labeling of the vertices and the se-

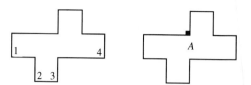

FIGURE **7.25**

the tiling indicated by Figure 7.26. With the configuration o[
labeled 3 in the slot now also forbidden, the final possibility c
slot with the corner labeled 4 quickly gives the tiling indicate(
7.27. All the possibilities have been examined. The hexomin(

FIGURE **7.26**

In Problem 7.16 we are asked to show that the hexomino in F
7.28 is 3-poic. We will only outline the backtracking that proves
With hindsight, it is seen that the labeling of the vertices and th(
lected slot, shown in Figure 7.28, are convenient. Due to the 180° r
tion symmetry of the hexomino, we need label only three corners.
configuration of the corner labeled 1 in the slot indicated in copy A le
to the tiling indicated by Figure 7.29. The configuration of the cor.
labeled 2 in the slot in copy A then leads to a tiling that is suggested
Figure 6.26; the configuration of the corner labeled 3 in the slot in co

FIGURE 7.29

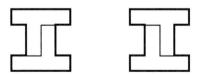

FIGURE 7.30

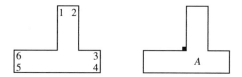

FIGURE 7.31

FIGURE 7.32

lected slot, shown in Figure 7.31, are convenient. We need to label all the corners due to the lack of symmetry of the hexomino. Digress from the natural order and show that the configuration formed by placing any of the corners labeled 2, 3, or 5 into the indicated slot in copy A leads to an impossibility. Then, placing the corner labeled 1 in the slot in copy A leads to the tiling indicated by Figure 7.32. Next, the configuration of the corner labeled 4 in the slot in copy A then leads to a tiling that is indicated by Figure 7.33. Finally, the configuration of the corner labeled 6 in the slot in copy A now leads to a tiling that is indicated by Figure 7.34. You should be able to relate the tilings indicated by Figures 7.33 and 7.34 with those indicated by Figures 6.26 and 6.25. The hexomino is tripoic.

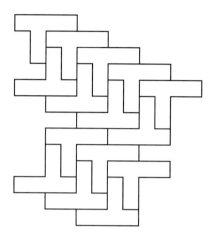

FIGURE 7.33

FIGURE 7.34

THE HEPTOMINOES

There are 108 heptominoes. Of all these, the maverick is shown in Figure 8.1. It's the one with a hole. With a set of the 108 heptominoes, we cannot cover any figure without a hole in it. If we choose to ignore the holey one, then we have a chance. We have 107 unholey heptominoes. Since $(7)(107)$ is a product of two primes, the only rectangle that can possibly be tiled with the 107 unholey heptominoes is a 7×107 rectangle.

FIGURE 8.1

Problem 8.1

Cover a 7×107 rectangle with the 107 unholey heptominoes.

Problem 8.2

Using all of the 108 heptominoes, finish covering three 11×23 rectangles that have had their central unit squares already covered.

Problem 8.3

Using all of the n-ominoes for which $1 \le n \le 7$, cover a 5×211 rectangle.

A solution to Problem 8.1 is indicated at the beginning of this chapter. All we have to do is unwind the figure, which essentially consists of one 7×7 square and four 7×25 rectangles and which contains each of the 107 unholey heptominoes. Just forming the set of heptominoes is a considerable task. Problem 8.2 is stated for those who would like to use all 108 heptominoes. For those who would like to use all of the heptominoes and yet not have holes in their figures, there is the possibility of considering the hole as a monomino. Well, if you are going to do that, then you might as well throw in all of the other polyominoes in between. This is what is done in stating our third problem. Only the most determined puzzle solvers will attempt to put the required 164 polyominoes together to solve Problem 8.3. It can be done.

Tiling with Heptominoes

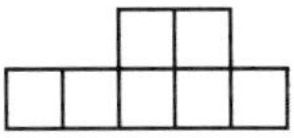

FIGURE **8.2**

FIGURE **8.3**

Problem 8.4

Tile a rectangle with the heptomino in Figure 8.2.

Problem 8.5

Tile a 14 × 14 square with the heptomino in Figure 8.3.

Problem 8.4 is very, very hard. Problem 8.5 is very hard. Twenty-eight copies of the heptomino in Figure 8.3 are required to cover a rectangle. A solution to Problem 8.5 that has a 90° rotation symmetry will be given later. Thirty-six copies of the same heptomino will also tile a 12 × 21 rectangle, if you want an additional puzzle.

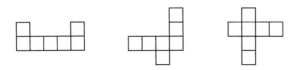

FIGURE **8.4**

Problem 8.6

Show that none of the three heptominoes in Figure 8.4 tiles the plane.

Problem 8.7

Show that the heptomino in Figure 8.5 tiles the plane in infinitely many ways but is monopoic.

FIGURE 8.5

So the holey heptomino is not the only heptomino that fails to tile the plane. The four heptominoes shown in Figures 8.1 and 8.4 are the first of the polyominoes that fail to tile the plane, and they are the only heptominoes that have this property. At the other extreme, there are ninety heptominoes that tile the plane in an infinite number of ways. Then, as we will see, fourteen of the heptominoes are polymorphic.

Problem 8.6 and part of Problem 8.7 require the method of backtracking. We will do the easy part of Problem 8.7, showing that there are an infinite number of tilings of the plane with the heptomino. As an example, we will show this in several related ways. An examination of the idea of stacking copies of an infinite strip will lead to the following observation: If a polygon tiles an infinite strip having parallel edges and if this strip has a symmetry that does not preserve the set of individual polygons tiling the strip, then the polygon tiles the plane in infinitely many ways.

The parallel edges of an infinite strip are exactly what is needed for stacking. Two curves are said to be parallel if you can slide one onto the other without any rotation. Such a slide is more formally called a translation and moves every point the same distance in the same direction. The two edges of the infinite strip indicated in Figure 8.6 are parallel, since we can easily slide the left edge to fit on top of the right edge. The strip having the highlighted edges in the figure and not yet tiled with the hexomino can be stacked in a unique way to cover the plane, once one strip is positioned. The stacking is made possible by the congruence under translation of the two edges. Conversely, the images of one edge under repeated application of the translation and its inverse form a partition of the plane into infinite strips with parallel edges.

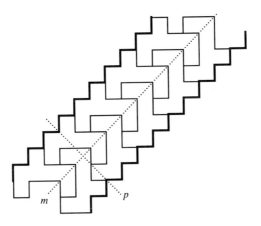

FIGURE 8.6

A translated copy of the strip tiled with hexominoes from Figure 8.6 can be stacked to the left of a given copy in ways that provide two non-congruent configurations. The second configuration is obtained from the first by translating the left copy "up one stair." Moving up one more stair brings us back to the first configuration. Moving up one stair describes the result of the translation along the line labeled m in the figure through the length of a diagonal of a unit square. This translation preserves the strip but does not preserve the set of individual hexomino tiles in the strip. In particular, under this translation, the image of the strip falls exactly on itself but the image of any particular hexomino tile partially overlaps that particular tile and two of its neighbors. In stacking the strips, we always have a choice of which of the two possible configurations to form when placing the next strip to the left of the previous choice. Of course, a similar choice holds for those strips stacked to the right. It follows that we can stack copies congruent under translation in infinitely many ways to get infinitely many tilings of the plane. In a sense, this is the most difficult case to discern because our eye so easily identifies copies that have the same aspect. The situation becomes more clear as we turn to other symmetries of the strip.

Another symmetry of the strip in Figure 8.6 that does not preserve the set of tiles in the strip is the 180° rotation about the point of intersection of the lines m and p in the figure. The rotated copy looks quite different to the eye. We can arbitrarily label one copy "up" and the rotated copy "down." In stacking copies of these strips to tile the plane, we have a choice at each stage of selecting the up copy or the down copy. We have infinitely many stackings. To make sure we don't have infinitely many copies of only a finite number of tilings, we can produce an infinite set without duplication in the following manner, which applies to all the related cases where we have two possibilities at each stage in the stacking. We start with an initial up copy. We let all copies to the left of this initial copy be down copies. Then each determination of all the copies to the right of the initial copy determines a different tiling of the plane. Since we have two choices for each of the infinitely many copies to the right, there is an infinite set of distinct tilings. (For those who do not want to make an infinite number of choices, the set of tilings formed by n adjacent up copies surrounded on both sides by down copies only will provide an infinite set of distinct tilings, if we let n run over the set of all the positive integers)

A reflection in a line is another possible type of symmetry for an infinite strip. Our strip is preserved by the symmetry of reflection along its midline, which is labeled m in Figure 8.6. This reflection does not preserve the set of hexominoes in the strip. Independent of the previous arguments, we again have two aspects for stacking and can produce an infinite number of tilings of the plane. Another line of symmetry is the line labeled p in the figure. Independent of all the arguments above, we can come to the same conclusion as before, namely that there are infinitely many tilings of the plane with the hexomino. Note that because of the abundance of symmetries in our example, the strips we get by the reflections are congruent under translation to either the strip in the figure or to this strip rotated 180°. This need not be the case in general. In general, you should not expect as many symmetries to be lurking around as there are in this example. One is all that is needed, however. Any symmetry of the strip that fails to be a symmetry of the set of individual tiles in this strip leads to an infinite number of tilings of the plane.

So far we have seen that the symmetries of an infinite strip might include translations, rotations, and reflections. Although it is by no means obvious, those who have studied symmetries of the plane know that there is only one more possibility for a symmetry of the infinite strip. It is possible that there is a translation along some line m followed by a reflection in the same line m such that the combination preserves the strip but not the set of tiles in the strip, although neither the translation alone nor the reflection alone has this property. (The formal name for such a symmetry is a "glide reflection." It can be shown that any figure can be made to coincide with a congruent figure by a translation (glide), a rotation, a reflection (flip), or a glide reflection.) Our example has too many symmetries for this to happen; see Figure 6.3 for an example, noting that neither of the two strips there can be folded upon itself although they are congruent to each other by sliding and turning over. If a polygon is polymorphic, then a symmetry of an infinite strip tiled with the polygon must preserve the set of polygons tiling the strip.

Figure 8.7 indicates the only tiling of the plane that can be made with directly congruent copies of the heptomino in Figure 8.5. Note that

FIGURE 8.7

this figure alone does not answer Problem 8.7. The figure shows that there is at least one tiling. You still need backtracking to prove that there is at most one tiling. It is, after all, certainly conceivable that there are more tilings than the one indicated by the given figure.

Polymorphic Heptominoes

Of all the n-ominoes with $n \leq 7$ that tile the plane, there is only one that requires copies that are not directly congruent to each other in order to tile the plane. This is the heptomino shown in Figure 8.8. Loosely speaking, you need to turn over some copies of the polyomino in order to tile the plane. Such a polygon is 0-poic. There are six monomorphic heptominoes. They are shown in Figures 8.8 and 8.9. Each of the heptominoes in Figure 8.9 has at least one line of symmetry, which means that for each of the heptominoes there is a line along which that heptomino can be folded in half. An m-morphic polygon having a line of symmetry is necessarily m-poic.

FIGURE **8.8**

Problem 8.8

Show that the heptomino in Figure 8.8 is monomorphic (1-morphic) but gymnopoic (0-poic).

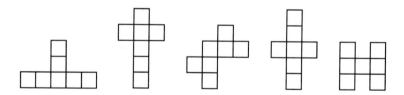

FIGURE **8.9**

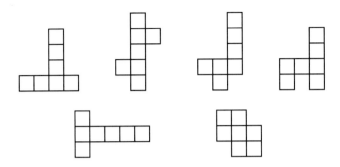

FIGURE 8.10

Problem 8.9

Show that each of the five heptominoes in Figure 8.9 is mono-morphic and monopoic.

Problem 8.10

Show that each of the six heptominoes in Figure 8.10 is dimor-phic and dipoic.

You should expect the last heptomino Z in Figure 8.10 to tile in the two ways that are indicated in Figures 6.25 and 6.26. They do, but there is a hitch. In this case these two tilings are the same; each is congruent to that indicated in Figure 8.11. The other tiling that is produced in the process of backtracking is indicated in Figure 8.12.

We have seen six monomorphic heptominoes and six dimorphic heptominoes. There are two more polymorphic heptominoes. Both are trimorphic.

Problem 8.11

Show that the heptomino in Figure 8.13 is trimorphic and dipoic.

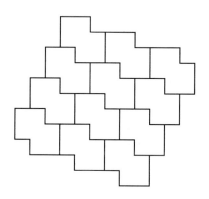

FIGURE **8.11**

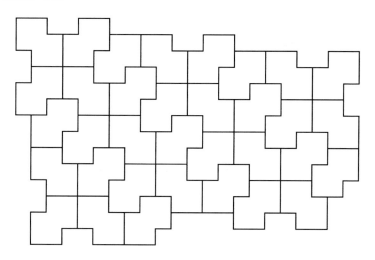

FIGURE **8.12**

FIGURE **8.13**

Problem 8.12

Show that the heptomino in Figure 8.14 is trimorphic and tripoic.

FIGURE **8.14**

FIGURE **8.15**

The tilings related to Problem 8.11 are indicated in Figure 8.15, and the tilings related to Problem 8.12 are indicated in Figure 8.16. You should be warned that the backtracking for Problem 8.11 is not easy.

More Remarkable Rectangles

Seventy-eight copies of the hexomino in Figure 8.17 are required to form the 21×26 rectangle in Figure 8.17. However, only seventy-six copies of this hexomino are required to form the 19×28 rectangle in Figure 8.18. This is the smallest rectangle that can be tiled with the hexomino.

The square in Figure 8.19 gives the promised symmetrical solution to Problem 8.5. Twenty-eight copies of this heptomino are needed to tile any rectangle.

A square can be cut into an odd number of congruent nonrectangular polygons. For example, Figure 8.20 shows twenty-five copies of an enneomino (9-omino) tiling a 15×15 square. Another example appeared as Problem 6.6 and was solved in Figure 6.27.

Problem 8.13

What is the smallest odd integer n such that a square can be cut into n congruent nonrectangular polygons?

Problem 8.14

Cut some square into fifteen congruent nonrectangular polyominoes.

Can a square be cut into three congruent nonrectangular polygons? Surely not, but there is no known proof. Note that the polygons in Problem 8.13 are not restricted to polyominoes. With or without the restriction, the answer to the the problem is unknown. However, the minimum is at most 15, as suggested by Problem 8.14. You may find a hint in Figure 3.9 that helps solve Problem 8.14.

FIGURE 8.16

FIGURE 8.17

Polyominoes that require $4k$ copies to form a rectangle were given by S. W. Golomb in 1988. One family is indicated in Figure 8.21, where we start with a T-tromino and add on $k - 1$ copies of a Z-shaped octomino in the manner shown in the figure. Three of the octomino Zs have been added to the tetromino in Figure 8.21 in order to produce a larger polyomino that requires sixteen copies in order to form a rectangle. This result gives a partial solution to the following two problems, whose solutions are unknown.

FIGURE 8.18

Problem 8.15

For which positive integers n does there exist some nonrectangular polygon such that n copies of the polygon form a rectangle?

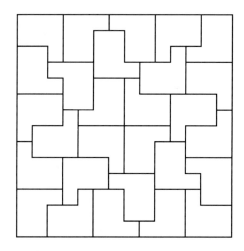

FIGURE 8.19

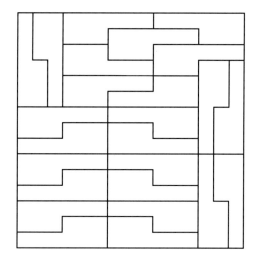

FIGURE 8.20

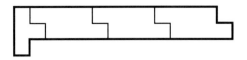

FIGURE **8.21**

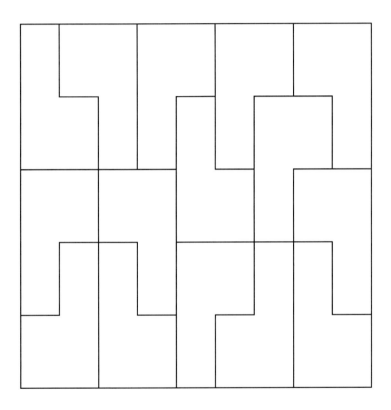

FIGURE **8.22**

Problem 8.16

For which positive integers n does there exist some polygon such that n is the minimum number of copies required to form a rectangle?

A square can be cut into fifteen congruent nonrectangular polygons. The solution in Figure 8.22 is a 45×45 square cut into fifteen congruent polyominoes having consecutive sides of lengths 18, 10, 9, 5, 9, and 5. This solution is obtained by appropriately stretching Figure 3.9.

POLYMORPHIC
POLYOMINOES

Unit squares may be placed along the twelve outer edges of the only holey heptomino to produce the six different holey octominoes. Altogether there are 369 octominoes. From the 363 unholey octominoes, we single out three for our first three problems. The straight octomino is pictured in Figure 9.1. Similarly, the straight n-omino consists of n unit squares attached end to end. You will need twenty-four copies of the octomino in Figure 9.2 if you intend to attack Problem 9.2. A sequence of problems that are analogous to Problem 9.3 will introduce polyominoes that are both p-morphic and p-poic, after a discussion of the first three problems.

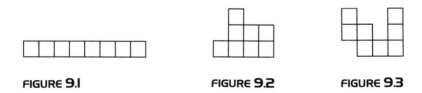

FIGURE 9.1   FIGURE 9.2   FIGURE 9.3

Problem 9.1

Find a fault-free tiling of a rectangle with the straight octomino.

Problem 9.2

Find the smallest rectangle tiled with the octomino in Figure 9.2.

Problem 9.3

Show that the octomino in Figure 9.3 is tetramorphic and tetrapoic.

Fault-Free Rectangles

In solving Problem 9.1, we set ourselves the more ambitious task of solving the more comprehensive Problem 9.4.

Problem 9.4

With $n > 1$, show that the smallest rectangle for which there is a fault-free tiling with the straight n-omino is a $(2n+1) \times 3n$ rectangle.

We begin the solution to Problem 9.4 with an argument that is very similar to that used in showing that the 5×6 rectangle is the smallest rectangle having a fault-free tiling with dominoes. In Figure 9.4, we suppose we are forming a fault-free tiling of an $h \times w$ rectangle with tiles that are copies of the straight n-omino. Now $h > n$, or otherwise all the tiles touching the left edge must have the same aspect and so form a fault. For now, we also suppose $h < 2n$. So there are tiles with a horizontal aspect and tiles with a vertical aspect touching the left edge. Since $h < 2n$, there must be exactly one vertical tile. This tile is denoted as region A in the figure; if the tile touches a corner of the rectangle, then ignore everything below the bottom arrow in the figure. The other $h - n$ tiles touching the left edge must be horizontal and are denoted by B in the figure. To the right of A there could be some vertical tiles but not so many as $n - 1$ of them. These vertical tiles, if there are any, in region C cannot proceed to the right edge of B without forming a vertical fault, if the tiling is to continue to the right, or forming a horizontal fault at the

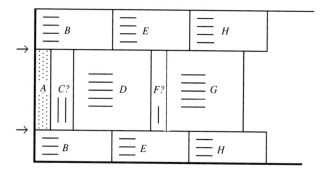

FIGURE 9.4

top arrow, if the tiling is complete. Therefore, we must have n horizontal tiles in position D to the right of C in the figure. These tiles extend beyond the right edge of region B. So, we must have $h - n$ horizontal tiles in position E in the figure. Now we could have some vertical tiles in the region F to the right of D. These vertical tiles, if there are any, in region F cannot proceed to the right edge of E without forming a vertical fault, if the tiling is to continue to the right, or forming a horizontal fault at the top arrow, if the tiling is complete. That sounds familiar. Therefore, we must have n horizontal tiles in position G to the right of F in the figure. We're in a cyclic rut. Since the process is to terminate in a rectangle at some stage, we must admit that we will end with a fault at the top arrow. So h is not less than $2n$.

The argument for the case $h = 2n$ is exactly the same as the argument above unless A is in a corner. See Figure 9.5. In this case, there is now room for adding some vertical tiles to the right of a pile of n horizontal tiles along the top as well as along the bottom. However, as in the previous argument, the vertical tiles, whether at the top or at the bottom, cannot extend as far to the right as the previously placed n horizontal tiles without forming a fault line. So $h > 2n$. By symmetry, we must also have $w > 2n$.

From our previous results, we know that if the $1 \times n$ rectangle is to tile an $h \times w$ rectangle the n must divide either h or w. The next integer after $2n$ that is divisible by n is $3n$; the next integer after $2n$ is $2n + 1$. Hence, the smallest possible rectangle that can have a fault-free tiling

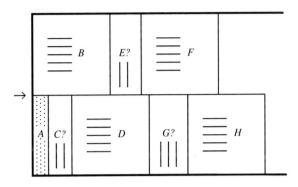

FIGURE 9.5

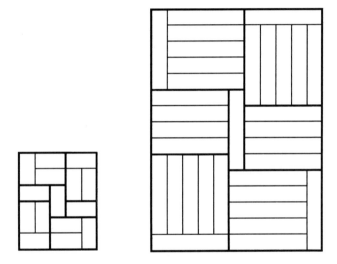

FIGURE 9.6

is a $(2n + 1) \times 3n$ rectangle. To finish the solution to Problem 9.4, we next produce a fault-free tiling for this rectangle.

The fault-free tiling of a 5×6 rectangle with dominoes from Figure 2.6 is repeated with highlights in Figure 9.6 to show how it is a special case of a general tiling. The figure actually shows the case for the straight pentominoes, but the analogous tiling for any straight n-omino is easily constructed. Check that for the general case the figure determines a

$(2n + 1) \times 3n$ rectangle. This completes the solution to Problem 9.1, as well as Problem 9.4. The tiling of the $(2n + 1) \times 3n$ rectangle is not unique. Figure 9.7 indicates how the tiling of the 5×6 rectangle with dominoes from Figure 2.8 is also a special case of a fault-free tiling of a $(2n + 1) \times 3n$ rectangle with straight n-ominoes.

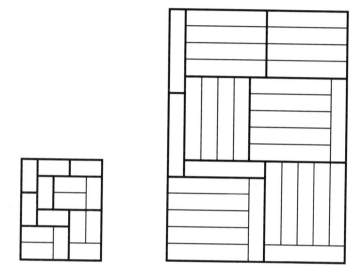

FIGURE 9.7

The tiling in Figure 9.8 gives a particularly pleasing solution to Problem 9.2 since the tiling is made up of four regions, each congruent to the other.

Conway's Criterion

The outstanding unsolved problem in tiling theory shows no sign of yielding any reasonable answer in the near future. The problem is: When does a given polygon tile the plane? A solution to the special case of polyominoes would be most welcome. In discussing a partial answer to the problem, we will encounter a result that is usually surprising if you

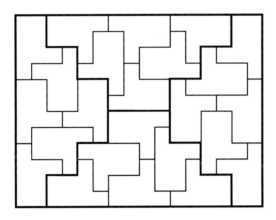

FIGURE **9.8**

have not thought about it. To exclude the problem because it does not involve polyominoes would be nearsighted and foolish. Any triangle is half a parallelogram and therefore easily tiles the plane. Surprisingly, any quadrilateral tiles the plane too. The quadrilateral in question need not even be convex, that is to say one vertex can be in the interior of the triangle formed by the other three vertices. Think about it.

Problem 9.5

Show that any quadrilateral tiles the plane.

A sufficient condition to determine whether a polygon tiles the plane is known to everyone interested in polyominoes beyond the puzzle problems. The condition has come to be known as the Conway criterion, after the English mathematician John H. Conway. The sufficiency depends on two facts, one of which is rather obvious and one that is not as obvious. The rather obvious fact is that if the opposite sides of a given hexagon are congruent and parallel then the hexagon tiles the plane. As in Figure 9.9, it is clear that copies of such a hexagon can be stacked end to end from a pair of opposite sides to form a strip with parallel edges and that copies of this strip can be stacked to tile the

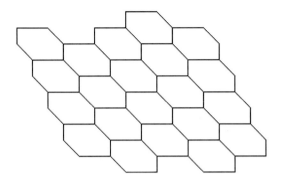

FIGURE **9.9**

plane. The less obvious fact, which is proved in the next paragraph, is the lemma: *The composite of one* 180° *rotation about a point followed by another* 180° *rotation about a second point results in a translation.* An immediate corollary of this lemma is what we will use. Specifically, we need the corollary: *A curve with a* 180° *rotation symmetry about point N is congruent and parallel to its image under a rotation of* 180° *about any point M.* This corollary follows because the 180° rotation about N leaves the curve fixed, since the curve is symmetric about N, and the second 180° rotation must then take the curve to a congruent and parallel curve since the composite of the two 180° rotations is a translation by the lemma. In Figure 9.10, for example, the 180° rotation about N fixes the S curve but interchanges A and B. The 180° rotation about M sends B to A', sends A to B', and sends the S curve with center N to some curve passing through N', the image of N under the 180° rotation about M. So the composite of the two 180° rotations sends A to A', B to B', and N to N'. The lemma tells us that this other curve is an S curve that is congruent and parallel to the S curve with center N. Informally, you might convince yourself that the lemma is to be expected by holding your pencil point on M in Figure 9.10 and rotating the page 180° about M to see that the S curve with center at N appears to take the place of the S curve with center at N'. Also, try this experiment with a curve that does not have a 180° rotation symmetry; you will find the image under the 180° rotation about M does not produce a parallel curve.

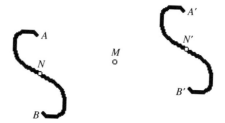

FIGURE 9.10

For a proof of the lemma, without loss of generality we let $N = (0,0)$ and $M = (a,b)$ in the cartesian plane. (If you are unfamiliar with coordinate geometry, just skip the rest of this paragraph.) We want to show that the $180°$ rotation about N followed by the $180°$ rotation about M results in a translation. We first show that a $180°$ rotation about any point (a,b) sends the point (x,y) to the point $(-x + 2a, -y + 2b)$. Suppose the point (x,y) is sent to (x',y') by the $180°$ rotation. The midpoint $((x + x')/2, (y + y')/2)$ of the two points (x,y) and (x',y') must be equal to the center (a,b) of the $180°$ rotation. Setting the respective coordinates equal, we get the equations $x' = -x + 2a$ and $y' = -y + 2b$, as claimed. For the special case $a = b = 0$, the rotation of $180°$ about the origin sends the point (x,y) to the point $(-x,-y)$. Now, as we just said, the $180°$ rotation about N sends (x,y) to $(-x,-y)$. The $180°$ rotation about M then sends the point $(-x,-y)$ to the point $(-(-x) + 2a, -(-y) + 2b)$. Hence, the composite of the two $180°$ rotations sends the point (x,y) to the point $(x + 2a, y + 2b)$. Since this composite adds the constant $2a$ to the abscissa of each point and adds the constant $2b$ to the ordinate of each point, the composite is a translation. Not that we need the fact here, but we cannot help noticing that this translation is the translation that moves each point twice the directed distance from N to M. This is worth noting, The composite of the $180°$ rotation about a point N followed by the $180°$ rotation about a point M is the translation that moves each point twice the directed distance from N to M. This elegant result more than finishes the proof of the lemma.

We now define a Conway hexagon, which need not even be a polygon. We suppose A, B, C, D, E, F are points not all on one line that are taken in cyclic order on the boundary curve of some region. Let $b(A, B)$ denote that part of the boundary curve connecting A and B that does not contain all the points A, B, C, D, E, F; with a similar notation adapted for the other "sides" of the "hexagon" having the "vertices" A, B, C, D, E, F. If $A = B$, then $b(A, B)$ is just a point, but there is no harm in that for what follows. Such a figure is a *Conway hexagon* if

1. there is a translation that takes A to E and takes B to D and
2. each of the four sides $b(B, C)$, $b(C, D)$, $b(E, F)$, and $b(F, A)$ has a $180°$ rotation symmetry.

A Conway hexagon is said to satisfy the *Conway criterion*. For emphasis, the theorem we want to prove is stated as our next problem.

Problem 9.6

Show that a Conway hexagon tiles the plane.

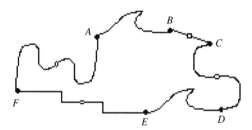

FIGURE 9.11

A rather complicated Conway hexagon is shown in Figure 9.11. It is much easier to see what is going on with this complicated example than with a polyomino, which has a boundary consisting of unit segments that have only two aspects. In addition, since the theory holds for these very generalized hexagons, there is no reason to confine ourselves to the polyominoes. The small empty circles in Figure 9.11 mark the centers of the $180°$ rotation symmetries for four sides of the the Conway hexagon.

Given a Conway hexagon with the notation above and as in Figure 9.11, let M be the midpoint of F and A. Since $b(F, A)$ is symmetric about

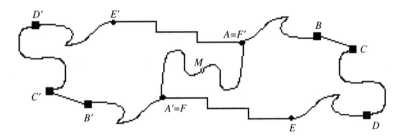

FIGURE 9.12

M, the Conway hexagon together with its image under a 180° rotation about M determine a new hexagon having vertices B, C, D, B', C', D', as in Figure 9.12. We will argue that this hexagon has opposite sides that are congruent and parallel. Since such a hexagon always tiles the plane by stacking, even when the sides are not line segments, then the half that is the Conway hexagon will also tile the plane. Using the expected notation related to the new hexagon, we know from the lemma that opposite sides $b(B, C)$ and $b(C', B')$ are congruent and parallel and that opposite sides $b(C, D)$ and $b(D', C')$ are congruent and parallel. (For example, the translation that takes B to C' and C to B' is the composite of the 180° rotation about the center of $b(B, C)$ followed by the 180° rotation about M.) This takes care of two pairs of opposite sides of the new hexagon. The remaining opposite sides are $b(D', B)$ and $b(B', D)$, each of which consists of three parts. We know that corresponding parts $b(A, B)$ and $b(E, D)$ are congruent and parallel and that corresponding parts $b(D', E')$ and $b(B', A')$ are congruent and parallel. By the lemma, we also know that corresponding parts $b(E', F')$ and $b(F, E)$ are congruent and parallel. Since $A = F'$ and $A' = F$, then the translation that sends B to D also sends A to E, sends E' to A', and sends D' to B'. So all in all, this marvelous translation takes the whole side $b(D', B)$ to the whole side $b(B', D)$. The opposite sides $b(D', B)$ and $b(B', D)$ of the hexagon B, C, D, B', C', D' are congruent and parallel, as desired. This finishes Problem 9.6.

Check that the heptomino in Figure 9.13 is marked to show that the figure is a Conway hexagon. Remember that a Conway hexagon need not be a hexagon in the usual sense of the word; the heptomino of the figure is actually a decagon. A different marking for the same heptomino is shown in Figure 9.14. Figure 9.15 corresponds to Figure 9.13 as Figure 9.12 corresponds to Figure 9.11. The tiling determined by Figure 9.13 is indicated in Figure 9.16 and is a repetition of the tiling at the top of Figure 8.16. The Conway hexagon of Figure 9.14 determines the middle tiling of Figure 8.16.

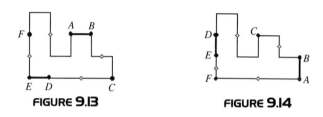

FIGURE 9.13   FIGURE 9.14

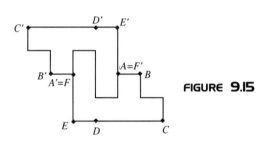

FIGURE 9.15

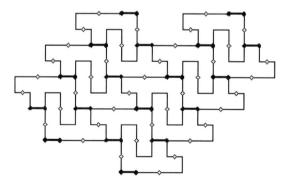

FIGURE 9.16

The vertices of a Conway hexagon must consist of at least three different points. Examples where the six vertices of a Conway hexagon are not distinct are shown in Figure 9.17. There is no restriction for a Conway hexagon that prevents the vertices *A*, *B*, *D*, *E* from being on one line. A special example of this is shown in Figure 9.18, which should lead to a solution for Problem 9.5.

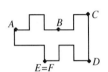 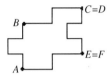

FIGURE 9.17

The Conway criterion is not a necessary condition for a polyomino to tile the plane. The heptominoes in Figures 8.5 and 8.8 tile the plane but do not satisfy the Conway criterion. However, knowing the Conway criterion does make life among the polyominoes much more enjoyable. You will find that it is a very useful tool. There is one warning that might be given about marking a curve as a Conway hexagon. Note that having $b(A, B)$ parallel to $b(D, E)$ is not sufficient; the translation that takes A to E must take B to D. Figure 9.19 shows an example of a polyomino that clearly does not tile the plane, even though $b(A, B)$ is congruent and parallel to $b(D, E)$ with each of each of the four sides $b(B, C)$, $b(C, D)$, $b(E, F)$, and $b(F, A)$ having a 180° rotation symmetry.

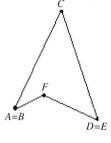   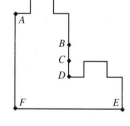

FIGURE 9.18   **FIGURE 9.19**

The Z's

By $Z(a, b, c, d)$ we will mean the polygon obtained by starting with an $(a + c) \times (b + d)$ rectangle and then removing the $b \times c$ rectangles from each of two diagonally opposite corners such that the resulting polygon has consecutive sides of lengths a, b, c, d, a, b, c, d. See Figure 9.20. In the construction, we must have $a > c$ or $d > b$ in order to have a polygon left after removing the corner rectangles. Note that $Z(a, b, c, d)$ is congruent to $Z(d, c, b, a)$. A $Z(a, b, c, d)$ is called a Z.

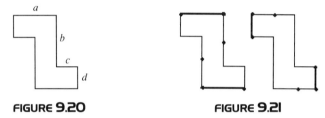

FIGURE 9.20   **FIGURE 9.21**

Any Z can be marked as a Conway hexagon in at least two ways, as shown in Figure 9.21. These two markings determine the tilings that are formed by stacking the Z's and that are indicated in Figures 6.25 and 6.26. Except when $a = d$ and $b = c$, these tilings are distinct. Of course, a Z might give rise to an infinite number of tilings, as does $Z(x + y, 2x, 2y, x + y)$. Check that out. On the other hand, dimorphic Z's are not hard to find. Trimorphic Z's are not as easy to find. At this time, there are five known families of trimorphic Z's. They are listed as follows.

1. $Z(a, a, c, c)$ with $a \neq c$, $a \neq 2c$, and $c \neq 2a$.
2. $Z(b + c, b, c, b + c)$ with $b \neq c$.
3. $Z(b + c, b, c, c)$ with $b \neq c$ and $b \neq 2c$.
4. $Z(2c, b, c, 2b)$ with $b \neq c$, $b \neq 2c$, $b \neq 3c$, $c \neq 2b$, and $c \neq 3b$.
5. $Z(6c, 3c, c, c)$.

An example from each of the five families is shown in Figures 9.22 and 9.23, where the tilings indicated are the ones that are not formed by stacking the Z's. Up to similarity, the fifth family contains only one member because $Z(6c, 3c, c, c)$ is similar to $Z(6, 3, 1, 1)$ for any positive number c.

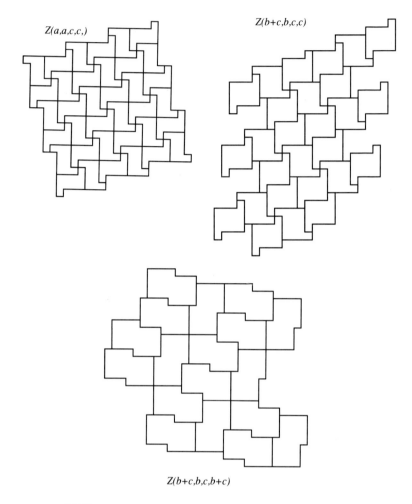

Z(a,a,c,c,)

Z(b+c,b,c,c)

Z(b+c,b,c,b+c)

FIGURE 9.22

Problem 9.7

Show that the Z's in Figure 9.22 are trimorphic and tripoic.

Problem 9.8

Show that the Z's in Figure 9.23 are trimorphic and dipoic.

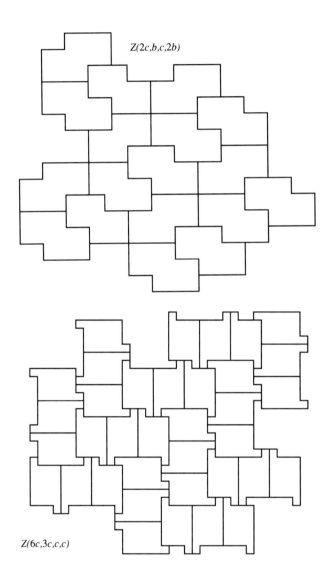

FIGURE **9.23**

Problem 9.9

Find some more trimorphic Z's.

The Z's are the key to all the polymorphic polyominoes that are presented below. Since a Z has a 180° rotation symmetry, we can easily bisect a Z into two congruent polyominoes. With planning and luck, copies of these halves can sometimes be put together to form other polymorphic polyominoes, in particular other polymorphic Z's.

Polypoic Polyominoes

The octomino in Figure 9.3 is the smallest polyomino that is tetramorphic. To solve Problem 9.3, backtracking is needed to show that there are no more than four tilings of the plane with the octomino. This will take very little time, unlike the many hours that it will take to do the backtracking required by some of the problems stated below. Two copies of the octomino form a trimorphic $Z(3, 1, 2, 3)$. A fourth tiling for the octomino is determined by a marking of the octomino as a Conway hexagon. Since each of these four tilings requires only copies that are directly congruent to each other, the octomino is also tetrapoic.

Problem 9.10

Show that the polyomino in Figure 9.24 is pentamorphic and pentapoic.

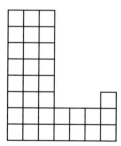

FIGURE 9.24

Two copies of the polyomino in Figure 9.24 form the dimorphic $Z(7,7,1,3)$, which can be seen inside the dimorphic $Z(15,6,3,7)$ formed by six copies in Figure 9.25. A fifth tiling that also uses only directly congruent copies is fun to find. Copies of the polyomino in Figure 9.26 form a dimorphic $Z(20,31,4,4)$ and a trimorphic $Z(24,24,4,4)$.

Problem 9.11

Show that the polyomino in Figure 9.26 is hexamorphic and hexapoic.

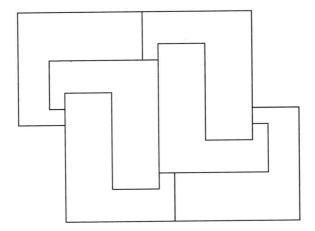

FIGURE 9.25

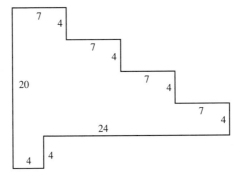

FIGURE 9.26

Problem 9.12

Show that the polyomino in Figure 9.27 is heptamorphic and heptapoic.

FIGURE 9.27   FIGURE 9.28

Two copies of the polyomino in Figure 9.27 form the trimorphic $Z(5, 4, 1, 1)$ of Figure 9.28. In addition, two copies form the dimorphic $Z(5, 5, 2, 1)$ of Figure 9.29 and the monomorphic polyomino of Figure 9.30. Finally, eight copies form the monomorphic polyomino in Figure 9.31. These polyominoes give rise to the seven tilings of the plane and require only directly congruent copies of the polyomino. Extensive figures are presented for this polyomino because $p = 7$ is as high as we will be able to go in illustrating p-poic polygons.

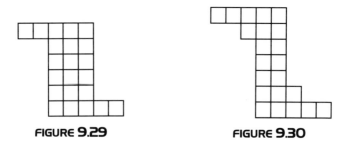

FIGURE 9.29   FIGURE 9.30

Problem 9.13

Show that the polyomino in Figure 9.32 is octamorphic and hexapoic.

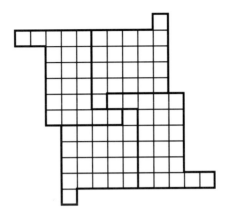

FIGURE 9.31

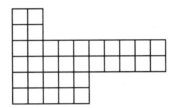

FIGURE 9.32

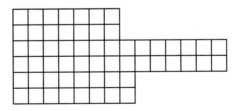

FIGURE 9.33

Problem 9.14

Show that the polyomino in Figure 9.33 is enneamorphic and heptapoic.

A Decamorphic Polygon

Problem 9.15

Show that the polyomino in Figure 9.34 is decamorphic and hexapoic.

Two copies of the polyomino in Figure 9.34 can be put together to form each of the three Z's in Figure 9.35. Nine of the ten tilings of the plane with this decamorphic polyomino are formed by various combinations of the these three Z's. Only the tenth tiling, which is indicated in Figure 9.36, contains no Z.

Infinitely many examples of m-morphic polyominoes are known for each n with $n < 10$. A figure similar to the one in Figure 9.34 is, of course also decamorphic. However, up to similarity, Figure 9.34 gives the only known decamorphic polygon. We have reached the following presently unsolved problems, where, as you might have guessed, a hendecamorphic polygon is an 11-morphic polygon. If you find a polygon that is hendecamorphic, dodecamorphic, triskaidecamorphic, or m-

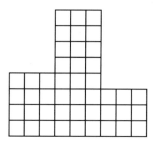

FIGURE 9.34

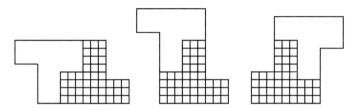

FIGURE 9.35

morphic for any m greater than 10, please communicate your result to the author.

Problem 9.16

Find another decamorphic polygon.

Problem 9.17

Find an octapoic polygon.

Problem 9.18

Find a p-poic polygon for some p such that $p > 8$.

Problem 9.19

Find a hendecamorphic polygon.

Problem 9.20

Find an m-morphic polygon for some m such that $m > 11$.

FIGURE 9.36

THE ANIMALS

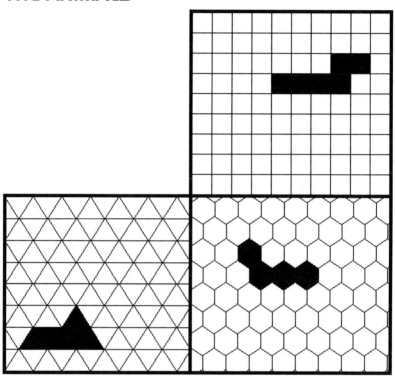

Snaky, the worm, and even the sphinx are all animals. Snaky is the hex-omino in the title figure and makes an interesting goal for the generalized tic-tac-toe game described earlier. Snaky is a square animal. The title figures also show the sphinx, which is a triangular animal, and the worm, which is a hexagonal animal. (Sorry, these are the standard names.) An *animal* is an edge-to-edge connected region that is a subset of the regular tiling of the plane by the equilateral triangle, the square, or the regular hexagon. The regular tilings are the three tilings that are indicated with the title; of the three regular polygons, only the hexagon is monomorphic. These are the only regular polygons that tile the plane. Why is that?

Animals start out as one cell and grow by the addition of similarly shaped cells. A formula for the number of n-celled triangular animals or for the number of n-celled hexagonal animals is as illusive as a formula for the number of n-ominoes. Practical applications for the animals range over the study of cell growth, organic chemistry, and computer design. Our introduction is solely for enjoyment. That is how people become mathematicians, by the way. They do mathematics because they enjoy it and eventually find themselves in a position to earn a living being paid to do what they like to do.

Generalizations of Polyominoes

The square animals have been called superdominoes and p-minoes, as well as polyominoes. The triangular animals are usually called *polyiamonds,* named by Thomas H. O'Beirne in his column in the British magazine *New Scientist.* Analogous to the naming of the polyominoes after dominoes, the choice of the name for the polyiamonds reflects the fact that a diamond consists of two equilateral triangles pasted edge-to-edge. Since O'Beirne says the i is part of the root of the word and not the prefix, we talk about n-iamonds when we want to specify that a polyiamond consists of n equilateral triangles. The hexagonal animals are usually called *polyhexes* and were named by the American mathematician and computer scientist David Klarner, who was among the first to study them and has also done a great deal of work on polyominoes. Of course, an n-hex consists of n regular hexagons. As expected, the Greek prefixes are used for naming the small animals. This does result in the name hexahex for a 6-hex. If that name upsets you, then you had better brace yourself for the next paragraph.

   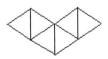

FIGURE 10.1

Some of the animals do have animal names. The lobster and the bat, which are hexiamonds, are shown in Figure 10.1. We might not expect the butterfly and the bee to be reptiles. As we will see when we meet them, they are not, but then it turns out that neither is the hexiamond that is called the snake. Is the snaky a reptile? Both the lobster and the bat are reptiles. Since we probably did not expect that all reptiles are animals, we may not find it too weird to realize that the wheelbarrow is an animal. Roger Penrose's loaded wheelbarrow, which is the 18-iamond shown in Figure 10.2, tiles the plane in a very interesting way.

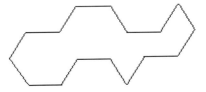

FIGURE 10.2

Problem 10.1

Show that the sphinx is a reptile.

Problem 10.2

Show that the lobster and the bat are reptiles.

Problem 10.3

Show that the loaded wheelbarrow tiles the plane.

We will return to the polyiamonds and the polyhexes after introducing some more generalizations of the polyominoes. First there are the so-called one-sided polyominoes. They are exactly what you think they are, polyominoes that you are not allowed to turn over. Six of the pentominoes are directly congruent to their images under a reflection; if you turn over any of the pentominoes denoted by the letters I, T, U, V, W, and X, then you get the same one-sided pentomino. However, none of the other six pentominoes is directly congruent to its image under a reflection; if you turn over any of the pentominoes denoted by the letters F, L, P, N, Y, and Z, then you get a different one-sided pentomino. There is therefore a total of eighteen one-sided pentominoes. (There are sixty one-sided hexominoes.) Arranging the eighteen one-sided pentominoes into rectangles provides many puzzles. One solution will be given for each of the following problems, which increase in difficulty with the order in which they are stated.

Problem 10.4

Cover a 9×10 rectangle with the eighteen one-sided pentominoes.

Problem 10.5

Cover a 6×15 rectangle with the eighteen one-sided pentominoes.

Problem 10.6

Cover a 5×18 rectangle with the eighteen one-sided pentominoes.

Problem 10.7

Cover a 3×30 rectangle with the eighteen one-sided pentominoes.

Stepping up a dimension is another way to generalize the idea of polyominoes. Set a unit cube on each unit square of a polyomino, gluing together the faces that touch, to get a solid polyomino. We will state two problems for the solid pentominoes. Problem 10.8 is said to have 3940 solutions. Then there are the polycubes. Polycubes include the solid polyominoes and are formed by gluing together cubes face-to-face. Polycubes are the complete three-dimensional analog to the two-dimensional polyominoes. Gluing together five cubes face-to-face gives a pentacube. There are twenty-nine pentacubes, where the mirror twins are counted separately. Since 29 and 5 are both primes, building boxes with all the pentacubes requires some adjustment. Only one more problem involving polycubes is stated here. You are on your own for finding solutions to these three-dimensional problems.

Problem 10.8

Construct a $3 \times 4 \times 5$ box with the twelve solid pentominoes.

Problem 10.9

Construct a $5 \times 5 \times 5$ cube from twenty-five copies of the solid Y-pentomino.

Problem 10.10

Omit the $1 \times 1 \times 5$ piece from the twenty-nine pentacubes and then construct two separate $2 \times 5 \times 7$ boxes with the remaining pentacubes.

FIGURE 10.3

FIGURE 10.4

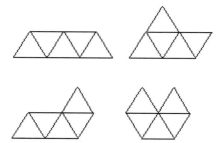

FIGURE 10.5

Polyiamonds

We return to the plane and the polyiamonds. There is one moniamond, one diamond, and one triamond. See Figure 10.3. The three tetriamonds are shown in Figure 10.4, and the four pentiamonds are shown in Figure 10.5. You ought to be able to formulate many questions and interesting puzzles that use these polyiamonds. Parallel your investigation along your study of the small polyominoes. Be inventive.

There are a dozen hexiamonds. They are given in Figure 10.6 with their original names. Note that the bat is called the chevron in the figure. Only the yacht and the sphinx cover twice as many triangles of one color as of the other color when placed on an alternately black and white colored regular tiling of the plane with the equilateral triangle. So the difference in the number of black triangles and white triangles in any figure to be covered with a set of the hexiamonds must be either 0 or 4. Such a figure that can be so covered is given in Figure 10.7.

Problem 10.11

Cover a 6 × 6 parallelogram with the twelve hexiamonds.

Problem 10.12

Which of the twelve hexiamonds tile the plane?

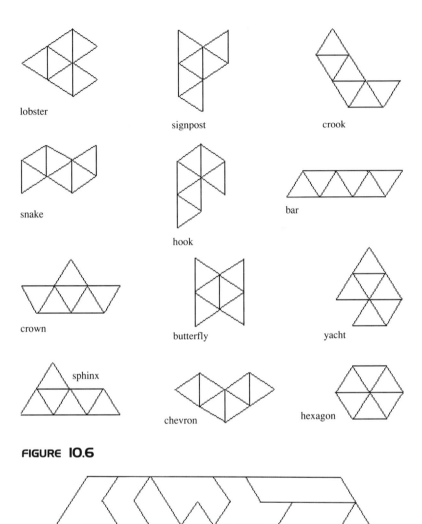

lobster

signpost

crook

snake

hook

bar

crown

butterfly

yacht

sphinx

chevron

hexagon

FIGURE 10.6

FIGURE 10.7

Problem 10.13

Which of the twelve hexiamonds are reptiles?

Problem 10.14

Find the twenty-four heptiamonds.

The only heptiamond that does not tile the plane is the V-heptiamond, which may be found at the lower right edge of Figure 10.8. A rather easy backtracking argument shows that the V-heptiamond does not tile the plane. A set of the twenty-four heptiamonds covers parallelograms that measure 3×28, 4×21, 6×14, and 7×12, as well as the symmetrical Figure 10.8, which has a subfigure that is similar to the whole.

Each of the sixty-six octiamonds tiles the plane. There are 160 enneiamonds. Twenty of the 159 enneiamonds without a hole and the unique heptiamond with a hole fail to tile the plane.

Copies of Figure 10.9, which consists of twelve copies of the wheelbarrow, can be stacked to tile the plane. Note that each of the twelve

FIGURE 10.8

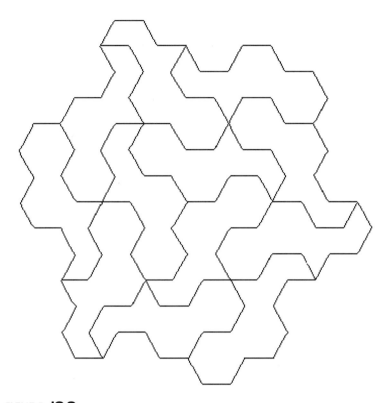

FIGURE 10.9

copies of the wheelbarrow in Figure 10.9 has a different aspect; Problem
10.3 makes an excellent jigsaw puzzle.

Problems 10.4 through 10.7 are solved in Figure 10.10, where each
rectangle is covered by the eighteen one-sided pentominoes.

There is one monohex, one dihex, but three trihexes. See Figure
10.11. The seven tetrahexes are shown in Figure 10.12 with their names.
There are twenty-two pentahexes and this is too many for nice puzzles.
Of the figures that you might want to try to cover with the seven tetrahexes, the "triangle" in Figure 10.13 is one of the first to pop up. However, it is impossible to cover this figure with the seven tetrahexes. A
backtracking proof starts with the observation that the propeller is limited to three positions, up to symmetry. The tower of Figure 10.14—o.k.,

FIGURE 10.10

so it's a lazy tower in the figure—can be covered with the seven tetra-hexes. Figure 10.15, which has three lines of symmetry as well as rotation symmetries of 120° and 240°, can also be covered with the seven tetrahexes.

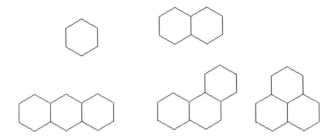

FIGURE 10.11

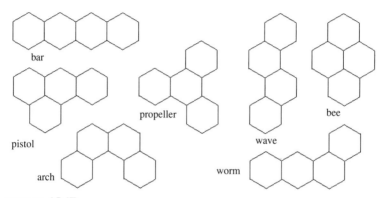

FIGURE 10.12

Problem 10.15

Show that the "triangle" in Figure 10.13 cannot be covered with the seven tetrahexes.

Problem 10.16

Cover the lazy tower in Figure 10.14 with the seven tetrahexes.

Problem 10.17

Cover the holey polyhex in Figure 10.15 with the seven tetra-hexes.

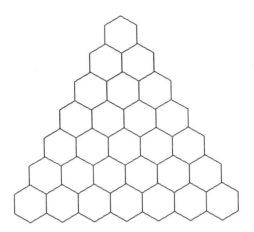

FIGURE 10.13

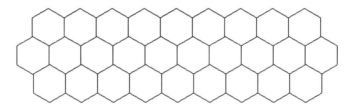

FIGURE 10.14

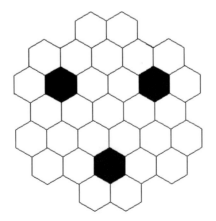

FIGURE 10.15

Problem 10.18

Which tetrahexes tile the plane?

Reptiles

The theory of the polyhex reptiles is very simple: A polyhex is never a reptile. On the other hand, arbitrarily many polyiamond reptiles can be created in a way analogous to that indicated in Figure 4.18 for polyomino reptiles. This time, begin with an equilateral triangle with sides of length $3m$ to produce $3m^2$-iamonds that are rep $9m^2$.

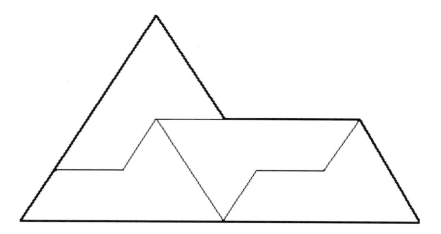

FIGURE 10.16

The hexiamonds that are reptiles are the sphinx, the lobster, the bat, and the bar. The sphinx is rep 4, as shown in Figure 10.16. The sphinx is the only known rep 4 pentagon. The lobster is rep 36, as shown in Figure 10.17. The bat or chevron is also rep 36 and has a tiling suggested by Figure 10.17. The bar is a parallelogram and therefore is rep 4.

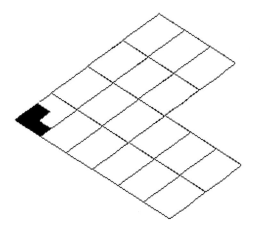

FIGURE IO.I7

Problem IO.I9

Show with four different tilings that the sphinx is rep 9.

Problem IO.2O

Show that each of the rep 4 reptiles in Figure 10.18 is also rep 9.

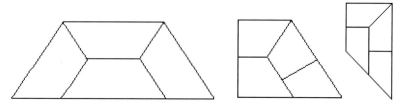

FIGURE IO.I8

Problem 10.21

Show that the rep 4 octiamond in Figure 10.19 is not rep 9.

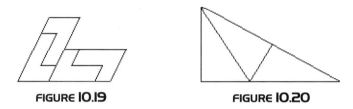

FIGURE 10.19   FIGURE 10.20

Ignoring the fact that not all reptiles are animals in the mathematical sense, we will take advantage of the name of this chapter to mention some of the things that are known and hint at some of the things that are unknown about reptiles. The altitude to the hypotenuse of an isosceles right triangle divides the triangle into two congruent triangles, each of which is similar to the original. Hence, an isosceles right triangle is rep 2. A 30°-60°-90° triangle is evidently rep 3, as seen from Figure 10.20. Any triangle is easily seen to be rep 4 by joining the midpoints of the sides of the triangle. The L-tromino is a special case of the reptile formed by removing a quadrant from any rectangle. Figure 10.18 gives the only three known examples of rep 4 quadrilaterals that are not parallelograms. Every known reptile tiles a parallelogram. Is there a reptile that does not tile a parallelogram? A parallelogram is the only known rep 7 polygon. For integer k with $k > 1$, a $1 \times \sqrt{k}$ parallelogram is rep k. You can see that the reptiles present many unsolved problems.

Figure 10.21 gives a solution to Problem 10.16, covering the lazy tower with the seven tetrahexes.

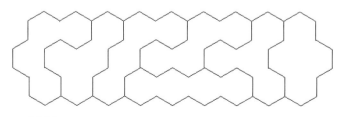

FIGURE 10.21

Some Numbers

We close with some numbers that were promised earlier.

n	# Unholey n-ominoes	# Holey n-ominoes	Total # n-ominoes
1	1	0	1
2	1	0	1
3	2	0	2
4	5	0	5
5	12	0	12
6	35	0	35
7	107	1	108
8	363	6	369
9	1,248	37	1,285
10	4,460	195	4,655
11	16,094	979	17,073
12	58,937	4,663	63,600
13	217,117	21,474	238,591
14	805,475	96,496	901,971
15	3,002,520	425,365	3,427,885

# n-iamonds	n	# n-hexes
1	1	1
1	2	1
1	3	3
3	4	7
4	5	22
12	6	82
24	7	333
66	8	1,448
160	9	6,572
448	10	30,490
1,186	11	143,552
3,334	12	683,101

Suggested Reading and References

Golomb's book is the classic reference to polyominoes. A revision is forthcoming from Princeton University Press. Although Golomb's two articles, cited below, are from a mathematical research journal, they are accessible to readers of this book and are worth investigating.

Solomon W. Golomb, *Polyominoes,* Scribner's, New York, 1965.

Solomon W. Golomb, Tiling with polyominoes, *Journal of Combinatorial Theory,* 1 (1966) 280–296.

Solomon W. Golomb, Tiling with sets of polyominoes, *Journal of Combinatorial Theory,* 9 (1970) 60–71.

Over the years, Martin Gardner's monthly column "Mathematical Games" from the magazine *Scientific American* has intrigued and stimulated many budding mathematicians, many professional mathematicians, and many more laypersons who just enjoy mathematics. Although any of these columns is worth reading, the following are particularly related to our topic:

Suggested Further Reading: October 1965 (Pentomino games), June 1967 (Polyhexes), September 1972 (Polycubes), July 1975 (Tiling), December 1975 (Tiling), April 1979 (Polyomino tic-tac-toe games), October 1979 (Packing squares), January 1977 (Advanced tiling)

Related Reading: May 1957, December 1957, November 1960, June 1961, November 1962, May 1963, December 1964, March 1965, July 1965, December 1966, March 1967, May 1973, August 1975.

The Mathematical Gardner (David A. Klarner, editor, Wadsworth, Belmont CA, 1981) is a collection dedicated to Martin Gardner and has many very interesting articles, including the following.

R. L. Graham, Fault-free tilings of rectangles, 120–126.

B. Grünbaum and G. C. Shephard, Some problems on plane tiling, 167–196.

C. J. Bouwkamp, Packing handed pentacubes, 234–242.

D. A. Klarner, My life among the polyominoes, 243–262.

The *Journal of Recreational Mathematics* contains more material on polyominoes than any other journal. You might investigate this journal for general enjoyment. As with any magazine, you may not be interested in every article, but then you may find enough in each issue to make it worth while to subscribe to the journal. (The journal is relatively inexpensive; contact Baywood Publishing Company, P.O. Box 337, Amityville NY 11701, USA.) This quarterly is one way to get involved in today's mathematics. There are always problems submitted by the readers. If you have access to a library that has the journal, you may wish to look up the problems numbered 381, 600, 881, 907, 929, 982, 1045, 1064, 1193, 1277, 1319, 1329, 1347, 1463, 1615, 1665, and 1791. The following list of articles from the *Journal of Recreational Mathematics* contains more suggested reading and references for much of the material that appears in this book.

Brian R. Barwell, Clever construction, 8 (1975–76) 130.

David Bird, The known world of octominoes, 8 (1975–76) 300–301.

C. J. Bouwkamp, Simultaneous 4 × 5 and 4 × 10 pentomino rectangles, 3 (1970) 125.

I-Ping Chu and Richard Johnsonbaugh, Tiling boards with trominoes, 18 (1985–86) 188–193.

Jenifer Haselgrove, Packing a square with Y-pentominoes, 7 (1974) 229.

David A. Klarner, Packing a rectangle with congruent n-ominoes, 7 (1969) 107–115.

Earl S. Kramer and Frits Göbel, Tiling rectangles with pairs of pentominoes, 16 (1983–84) 198–206.

Andy Liu, Pentomino problems, 15 (1982–83) 8–13.

Joseph S. Madachy, Pentominoes—Some solved and unsolved problems, 2 (1969) 181–188.

Jean Meeus, Some polyomino and polyamond problems, 6 (1973) 215–220.

Jean Meeus, The smallest U-N square, 18 (1985–86) 8.

Jean Meeus and P. J. Torbijn, The 30/36 problem with pentominoes/hexiamonds, 10 (1977–78) 260–266.

Robert L. Patton and Joseph S. Madachy, The twenty problem—a limited solution, 3 (1970) 207–213.

Wade E. Philpott, Polyomino and polyiamond problems, Part I, 10 (1977–78) 2–14.

Wade E. Philpott, Polyomino and polyiamond problems, Part II, 10 (1977–78) 98–105.

Wade E. Philpott, The double-double pentomino problem, 14 (1981–82) 61.

P. J. Torbijn, The unknown world of octominoes, 1 (1974) 1–7.

For further reading on polymorphic polyominoes, consider:

A. Fontaine and G. E. Martin, Polymorphic polyominoes, *Mathematics Magazine,* 57 (1984) 275–283.

The following are listed for historical reference:

Henry Ernest Dudeney, *The Canterbury Puzzles and Other Curious Problems,* Dover (Reprint of book published in 1907), New York, 1958.

W. Stead, Dissection, *The Fairy Chess Review,* 9 (1954) 2–4.

S. W. Golomb, Checker boards and polyominoes, *American Mathematical Monthly,* 61 (1954) 675–682.

T. H. O'Beirne, Puzzles and paradoxes No. 44: Pentominoes and hexiamonds, *New Scientist,* 259 (November 2, 1961) 316–317.

Polyominoes are available commercially. From time to time, *Hexed,* a plastic set of the pentominoes, appears in toy stores. Inexpensive sets of plastic pentominoes are available from Creative Publications (800-642-0822; 5040 West 111th Street, Oak Lawn IL 60453). Kadon Enterprises, Inc. has elegant sets of polyominoes among their many game offerings (301-437-2163; 1227 Lorene Drive, Suite 16, Pasadena, MD 21122).

INDEX